English for Elevator

电梯专业英语

▶ 肖伟平　夏龙军　主　编
▶ 张　书　吕晓娟　副主编

·北京·

本书共11章，每章设置内容描述、知识准备、章节内容、自我评价和拓展阅读五个部分，重点介绍电梯的各个环节，涉及电梯（扶梯）结构，电梯安装及维护等多方面内容，同时也介绍了金属加工、焊接、工程制图等与电梯技术相关的专业知识，力求达到贴近工作与生产实际。本书采用大量的图片穿插说明，减少文字篇幅以降低阅读难度。每个章节的拓展阅读，具有一定的趣味性和实用性。

本书可供从事电梯生产、安装及维护技术人员阅读和参考，也可作为高等院校电梯专业及相关从业人员技术培训教材和参考用书。

图书在版编目（CIP）数据

电梯专业英语/肖伟平，夏龙军主编. —北京：化学工业出版社，2012.6（2025.6重印）
ISBN 978-7-122-14271-9

Ⅰ. 电… Ⅱ. ①肖…②夏… Ⅲ. 电梯-英语 Ⅳ. H31

中国版本图书馆CIP数据核字（2012）第094461号

责任编辑：廉　静　　文字编辑：林　丹
责任校对：顾淑云　　装帧设计：王晓宇

出版发行：化学工业出版社（北京市东城区青年湖南街13号　邮政编码100011）
印　　装：北京科印技术咨询服务有限公司数码印刷分部
710mm×1000mm　1/16　印张15　字数294千字　2025年6月北京第1版第9次印刷

购书咨询：010-64518888　　售后服务：010-64518899
网　　址：http://www.cip.com.cn
凡购买本书，如有缺损质量问题，本社销售中心负责调换。

定　　价：41.00元

前言

Foreword

我国电梯行业的蓬勃发展需要大量的技术人才，无论是生产制造、采购销售还是安装维护及保养等，同时我国电梯业也逐渐与国际接轨，很多电梯企业正在不断地吸收和引进国外的先进理念与先进技术，这对相关人才及技术等的要求提出了更高的挑战，电梯从业人员的英语素质亟待提高。只有掌握一定的专业英语，才能与国外电梯科研及技术人员进行相关的技术交流或贸易，更好地学习和掌握国外先进技术。中山职业技术学院作为国内率先成立电梯专业的院校，在人才培养、实训基地等方面做了大量的探索和实践。本书旨在提高电梯从业人员、高等院校电梯专业学生专业英语的阅读能力，使读者能扩充知识面，为专业学习服务，提高读者的综合素质和未来继续学习及发展的能力。为读者将来及时学习国内外电梯专业领域的新知识，了解行业发展新动向打下坚实基础。

全书共分11章，每章设置内容描述、知识准备、章节内容、自我评价和拓展阅读五个部分，重点介绍电梯的各个环节，涉及电梯（扶梯）结构，电梯安装及维护等多方面内容，同时也介绍了金属加工、焊接、工程制图等与电梯技术相关的专业知识，力求达到贴近工作与生产实际。本书采用大量的图片穿插说明，减少文字篇幅以降低阅读难度。每个章节的拓展阅读，具有一定实用性。本书内容通俗易懂、语言简练，趣味性强，选材新颖，覆盖面较广，可供从事电梯生产、安装及维护技术人员阅读和参考，也可作为高等院校电梯专业及相关从业人员技术培训教材和参考用书。

本书由肖伟平、夏龙军主编，张书、吕晓娟副主编，肖伟平负责全书统稿。具体编写分工如下：肖伟平编写第1至第3章，夏龙军编写第4至第8章，潘斌、屈省源编写第9章，张书编写第10章，吕晓娟编写第11章。

本书在编写过程中参考了相关的文献资料，同时得到了专业教研室各位老师的大力支持，并提出了许多宝贵意见，使本书得以顺利完成，在此一并深表感谢。由于本书与以往的专业英语教材相比，可以直接借鉴的资料不多，另外编写时间和水平均有限，书中不妥之处在所难免，敬请广大专家和读者批评指正。

编者

2012年5月

目录 Catalogue

目录 Catalogue

Chapter 1

Structure of Elevator Ⅰ
电梯结构 Ⅰ

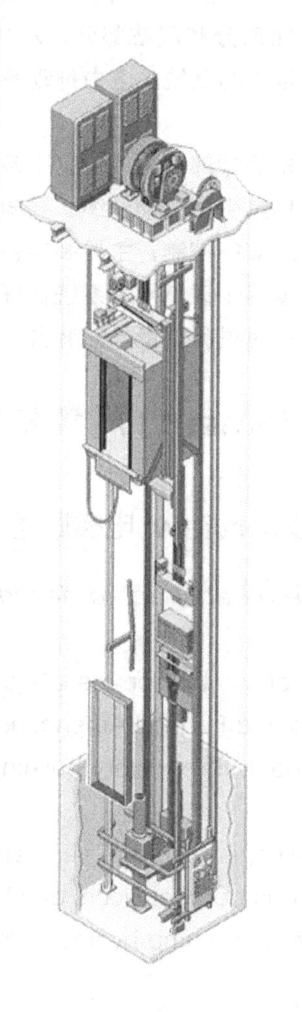

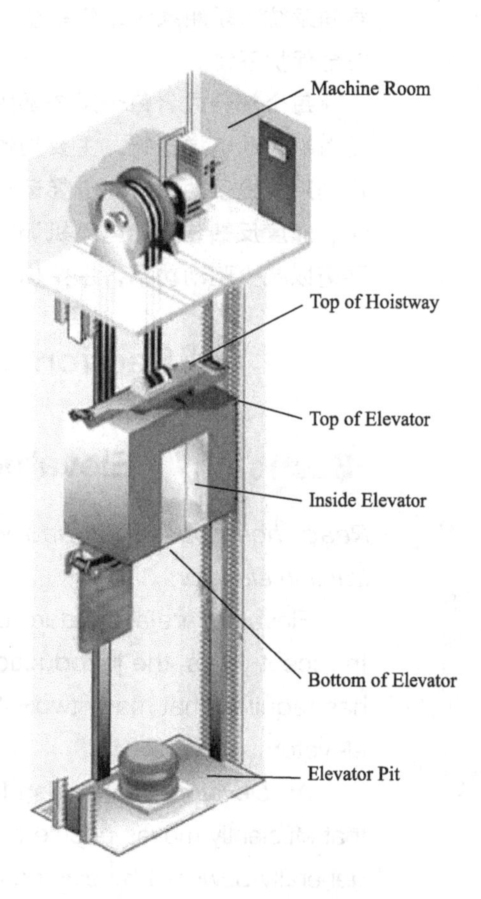

【Content Description】/【内容描述】

电梯是机电技术高度结合的特种设备，了解电梯的结构是制造、安装、使用和维护的前提。根据不同的标准，电梯有不同的种类，本书以目前使用最多的曳引型电梯为例，通过大量插图，对电梯进行介绍。

本章共设置了三节，电梯概述，曳引系统和重量平衡系统。通过本章的学习，了解电梯四大空间和八大系统的英语表达，掌握曳引系统、电力拖动系统和重量平衡系统中曳引机、制动器、对重和钢丝绳等常见零部件的英语词汇，能够翻译和看懂一般的电梯文章。

【Related Knowledge】/【知识准备】

典型曳引电梯的结构可以概括为四大空间八大系统：从位置或空间上可以分为机房部分、井道－底坑部分、轿厢部分和层站部分，从功能上分为曳引系统、导向系统、轿厢系统、门系统、重量平衡系统、电力拖动系统、电气控制系统和安全保护系统。

每个部分或者每个系统都由众多的零部件构成。曳引系统主要是输出与传递动力、驱动电梯运行，主要构件包括曳引机、钢丝绳、导向轮、反绳轮等；电力拖动系统提供动力、对电梯运行速度实行控制，主要装置包括曳引电机、供电系统、速度反馈装置、电动机调速装置等；重量平衡系统保证曳引轮两端重量接近平衡以减少驱动功率，主要构件包括对重和重量补偿装置。

【Section Implement】/【章节内容】

Section 1 Elevator Overview 电梯概述

Read these passages and write an abstract to describe the structure of traction elevator.

Elevators are a standard part of any tall commercial or residential building. In recent years, the introduction of the Federal Americans with Disabilities Act has required that many two-story and three-story buildings be retrofitted with elevators.

An elevator, lift in British English, is a type of vertical transport equipment that efficiently moves people or goods between floors of a building. Elevators are generally powered by electric motors or pump hydraulic fluid. In the application of vertical transportation systems, a major decision is which drive system to use, hydraulic or traction? Each type has characteristics which makes it particularly well suited for a specific application. In general, hydraulic elevators are suitable for low-rise buildings (up to 6 floors) whereas, the roped elevators (or "traction elevators" as Figure 1-1 shows) are best suited to higher buildings.

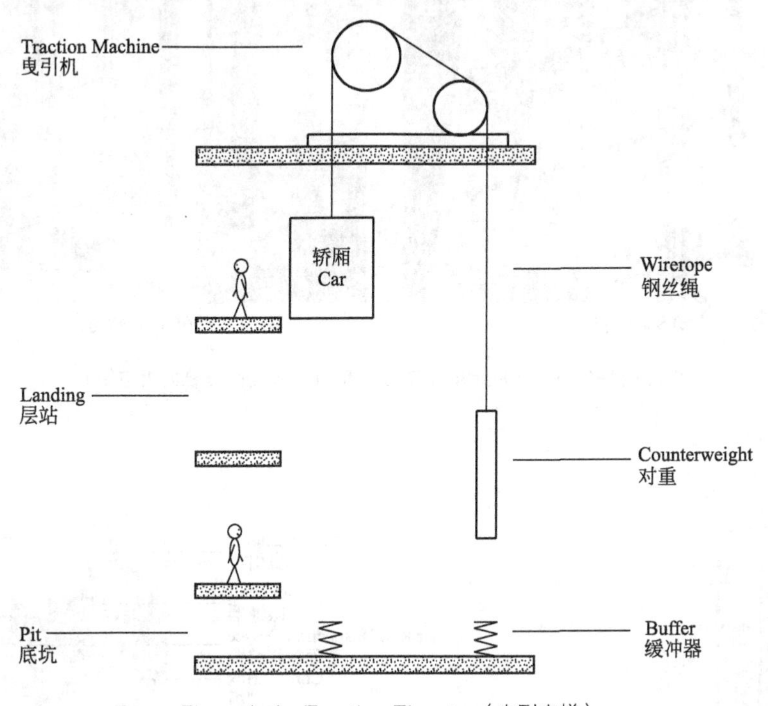

Figure1-1 Traction Elevator（电引电梯）

Traction elevators are the most popular type nowadays and are driven by the traction between the suspension ropes and the drive sheaves.

Elevators themselves are simple devices, and the basic lifting systems have not changed much in over 50 years. In space, elevators can be considered to be composed of four parts: machine room, shaft & pit, car and landing (as Figure 1-2 shows).

(a) Machine Room(机房)

(b) Landing (层站)

Figure1-2　Four Parts of Space for Elevator（电梯四大空间）

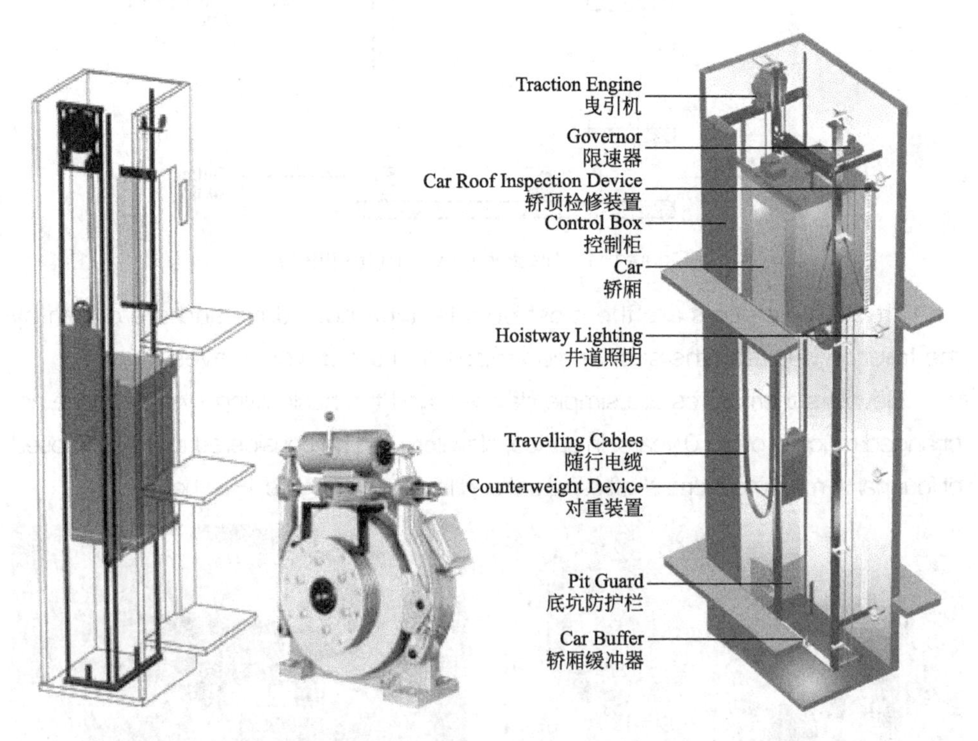

Figure1-3　Machine-Room-Less (MRL) Elevator（无机房电梯）

The machine-room-less elevator (as Figure 1-3 shows) is the result of technological advancements. These newly designed permanent magnet motors (PMM) allow the manufacturers to locate the machines in the hoistway overhead, thus eliminating the need for a machine room over the hoistway. This design has been utilized for at least 15 years and is becoming the standard product for low to low-mid rise buildings. It was first introduced to the U.S. market by KONE.

According to their function, elevator is made up of eight systems: traction system, guide system, car system, door system, weight balance system, electrical drive system, electrical control system and safety protection system (as Figure 1-4 shows).

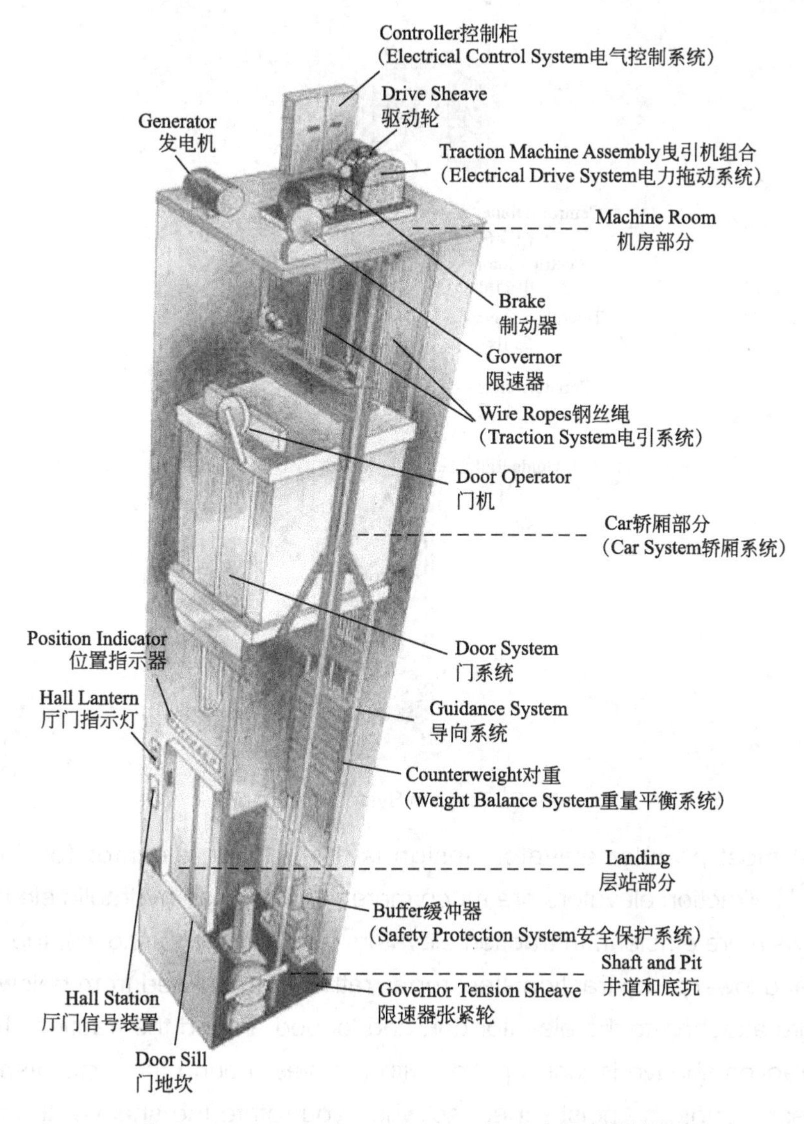

Figure1-4 Traction Elevator Overview(曳引电梯总览)

Notes and Expressions

1. commercial or residential building 商业或住宅建筑
2. retrofit ['retrəufit] *vt.* 改进；更新；改装 *n.* 式样翻新
3. PMM（permanent magnet motor）永磁电机
4. MRL（Machine-Room-Less）elevator 无机房电梯
5. traction ['trækʃən] *n.* 曳引；牵引

Section 2 Traction System 曳引系统

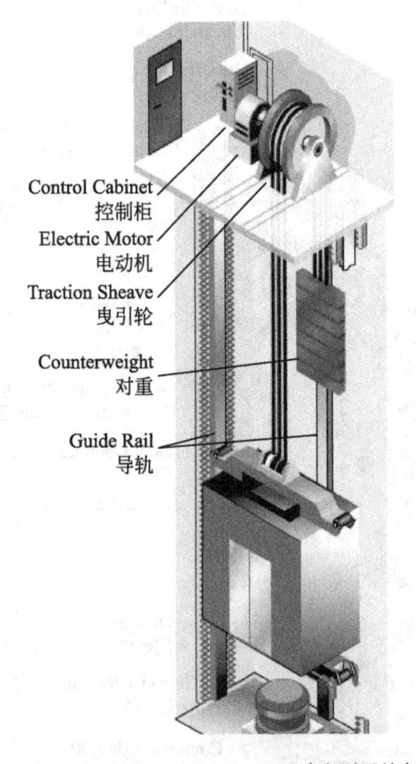

Figure1-5 Traction System（曳引系统）

The most popular elevator design is the traction elevator (or "roped elevator"). Traction elevators are much more versatile than hydraulic elevators, as well as more efficient. In traction elevators (as Figure 1-5 shows), the car is raised and lowered by traction steel ropes rather than pushed from below. The ropes are attached to the elevator car, and looped around the traction sheave (3). A traction sheave is just a pulley with grooves around the circumference. The sheave grips the hoist ropes, so when you rotate the sheave, the ropes move too.

The sheave is connected to an electric motor (2). When the motor turns one way, the sheave raises the elevator; when the motor turns the other way, the sheave lowers the elevator. In gearless elevators, the motor rotates the sheaves directly. In geared elevators, the motor turns a gear train that rotates the sheave. Typically, the sheave, the motor and the control cabinet(1) are all housed in a machine room above the elevator shaft.

The ropes that lift the car are also connected to a counterweight (4), which hangs on the other side of the sheave. The counterweight weighs about the same as the car filled to 40-percent capacity. In other words, when the car is 40 percent full, the counterweight and the car are perfectly balanced.

The purpose of this balance is to conserve energy. With equal loads on each side of the sheave, it only takes a little bit of force to tip the balance one way or the other. Basically, the motor only has to overcome friction — the weight on the other side does most of the work. To put it another way, the balance maintains a near constant potential energy level in the system as a whole. Using up the potential energy in the elevator car (letting it descend to the ground) builds up the potential energy in the weight (the weight rises to the top of the shaft). The same thing happens in reverse when the elevator goes up. The system is just like a see-saw that has an equally heavy kid on each end.

Both the elevator car and the counterweight ride on guide rails (5) along the sides of the elevator shaft. The rails keep the car and counterweight from swaying back and forth, and they also work with the safety system to stop the car in an emergency.

▪ Traction Machine and Brake 曳引机和制动器

Elevators with geared traction machine (as Figure 1-6 shows) are normally used on both passenger and freight elevators with rated speed up to 350 ft/min (≈ 1.78m/s). In a few cases, they have been used for speeds as high as 500 ft/min (≈ 2.54m/s).

Gearless machine (as shown in Figure 1-7) are based on the permanent magnet technology with high efficiency and have an energy saving of about 60%. They deliver the low speed, high torque performance which completely eliminated the gear box with less space required and totally smooth ride comfort travel.

Gearless traction elevators can reach speeds of up to 2000 ft/min (10 m/s), or even higher. The components are the same as previously described for geared units except for the driving machine.

Figure1-6　Geared Traction Machine（齿轮曳引机）

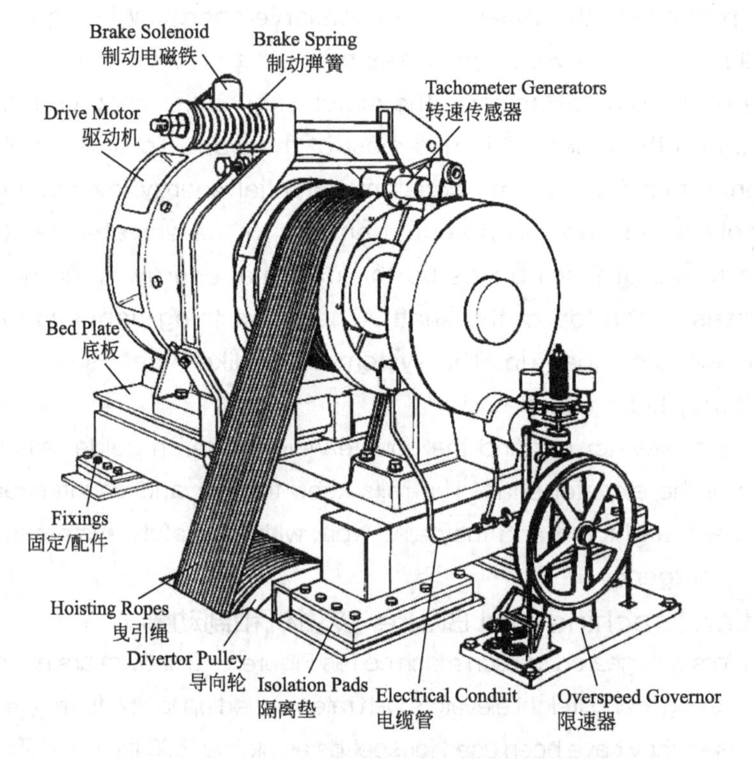

Figure1-7　Gearless Traction Machine（无齿曳引机）

The most common elevator brake is made up of a compressive spring assembly, brake shoes with linings, and a solenoid assembly. When the solenoid is not energized, the spring forces the brake shoes to grip the brake drum and induce a braking torque. The magnet can exert a horizontal force for the break release. This can be done directly on one of the operating arms or through a linkage system. The break is pulled away from the shaft and the velocity of the elevator is resumed.

In order to improve the stopping ability, a material with a high coefficient of friction is used within the breaks, such as zinc bonded asbestos. A material with too high a coefficient of friction can result in a jerky motion of the car. This material must be chosen carefully.

For both geared and gearless, the brake is released electrically and applied when electric power is removed. This brake is usually an external drum type and is actuated by spring force and held open electrically, a power failure will cause the brake to engage and prevent the elevator from falling. The drawings in Figure1–8 illustrate typical elevator brakes.

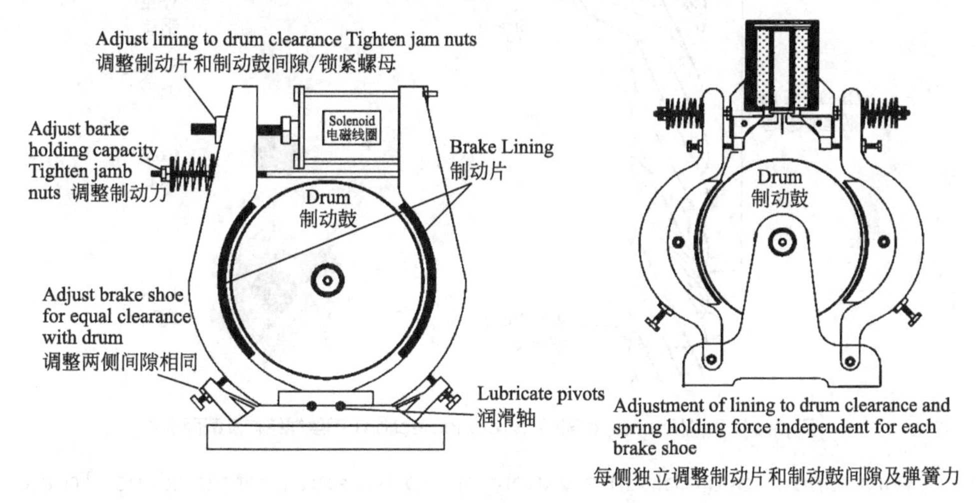

Figure1–8 Elevator Brake（电梯制动器）

▪ Traction Wire Rope 曳引钢丝绳

Due to its construction and the structure consisting of many individual steel wires, steel wire rope offers advantages that clearly qualify it for use on elevators. Its benefits are its redundancy and the capacity to identify the possibility of the end of service life or (preferably) the correct time for discarding the rope before its condition becomes dangerous by means of externally visible criteria such as wire breakages.

Wire rope (as Figure 1–9 shows) is made of wire strands and a core. The center wire is a round shaped wire used as the body member. Around this body member a group of wires are helically laid to form a strand. The strands are supported by the core, thus making up what we refer to as the wire rope diameter, which is utilized in manufacturing.

The greater the number of wires in a strand the more flexibility in the wire rope. The lower or less amount of wires in a strand the stiffer the wire rope. The center core may be made with a polypropylene Fiber Center (FC) or with a steel Independent Wire Rope Center (IWRC).

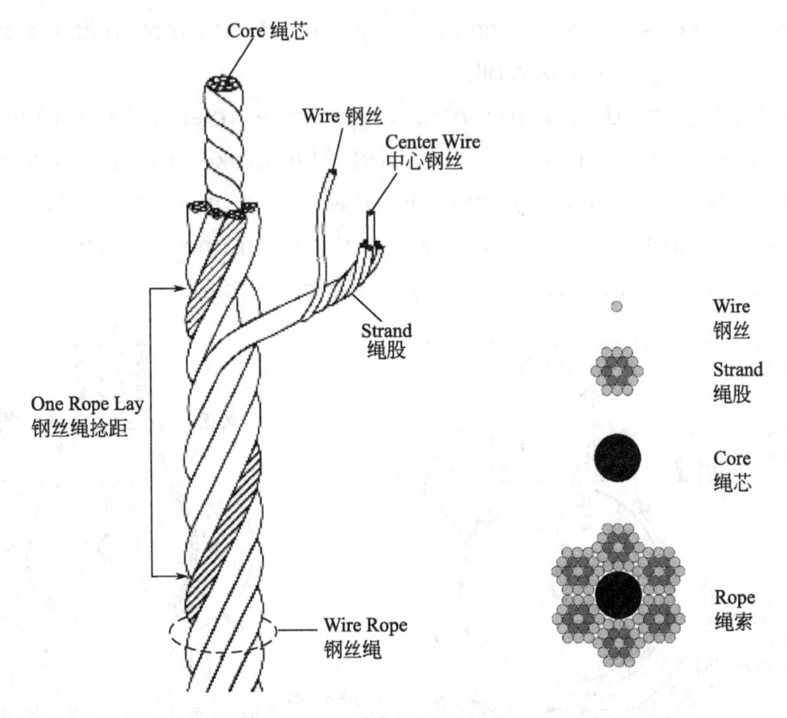

Figure1–9 Wire Rope Structure and Section（钢丝绳结构及截面）

Wire rope is available in a variety of grades and configurations. To the layman, the critical factors in selecting a rope are breaking strength and diameter (measurement methods see Figure 1–10).

An adequate factor of safety is crucial in wire rope use. For hoist rope of dumbwaiter, the recommended safety factor is 10 : 1. In other words, if the load weighs one ton, the wire rope used must have a minimum ultimate breaking strength of ten tons.

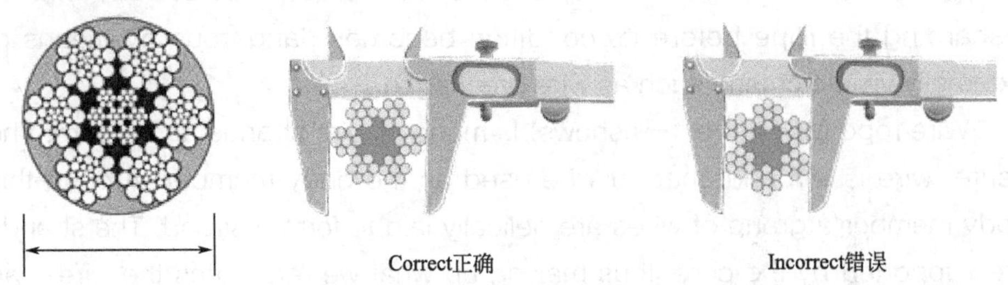

Figure1–10 Wire Rope Diameter Measurement Method（钢丝绳直径测量方法）

Rope diameter is important for compatibility with rigging hardware. In particular, the wire rope must seat properly in the sheaves to ensure freedom of movement without undue wear to rope or sheave (as Figure 1-11 shows).

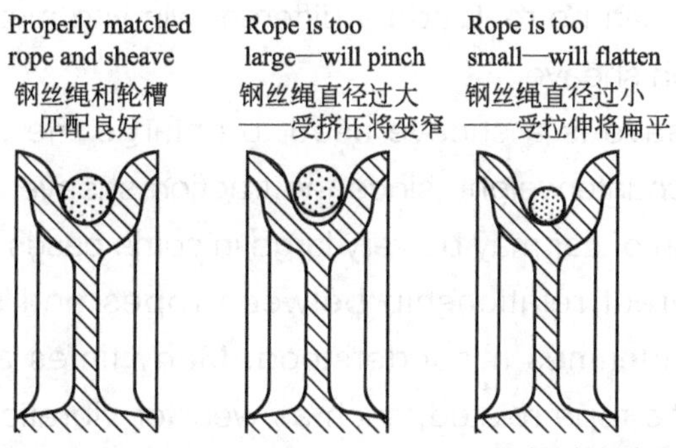

Figure1-11 Wire Rope Diameter and Sheave（钢丝绳直径与轮槽）

▪ Traction & diversion & deflector sheave 曳引轮/反绳轮/导向轮

Traction sheave is always driven by the drive machine and engaging the hoisting ropes. For gearless elevator, the traction sheave (as Figure 1-12 shows) is connected directly to the shaft of the traction motor, and the motor rotation (speed) is transmitted directly to the traction sheave without any intermediate gearing. For geared elevator, motor speed is reduced by 1/10th using a speed reducer equipped with worm or helical gears, and transmitted to the traction sheave of the traction machine.

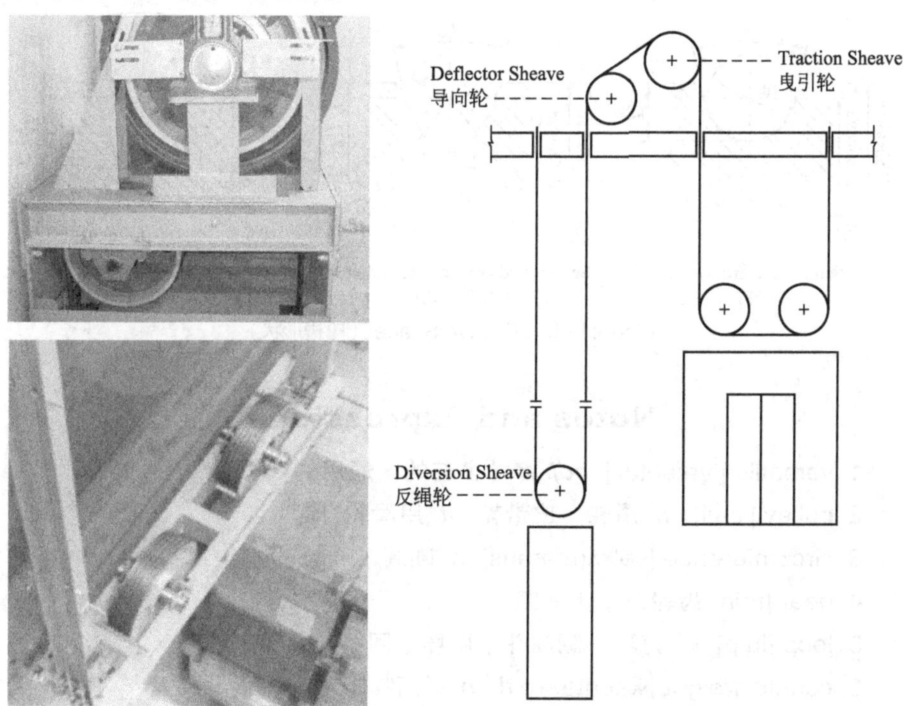

Figure1-12 Traction & Diversion & Deflector Sheave（曳引轮/反绳轮/导向轮）

Diversion sheave is a kind of fix pulley, usually locates in frame of car or counterweight, but they can be also found under car and on the wall of shaft. Most traction ratio can be realized by different winding method and a certain number of diversion sheave.

Deflector sheave is a sheave used to enlarge the distance between elevator car and counterweight, since the traction sheave diameter is limited while the dimension of car may be very large in some cases.

There is a direct relationship between ropes and sheaves, yet this relationship is sometimes misunderstood. Many times a problem occurs which appear to be rope related, such as wear or vibration, and the natural assumption of the cause is a manufacturing defect in the wire rope. More often than not, the actual cause of the problem is the sheave. Poor rope performance is the first indication of a sheave problem because a sheave problem is always conveyed to the ropes. It is important to remember that as the rope seats itself into the groove, the groove profile will be "molded" to the rope diameter.

Traction force used in elevator is affected by shape of sheave groove, material and surface quality. The main groove shapes are shown in figure1-13.

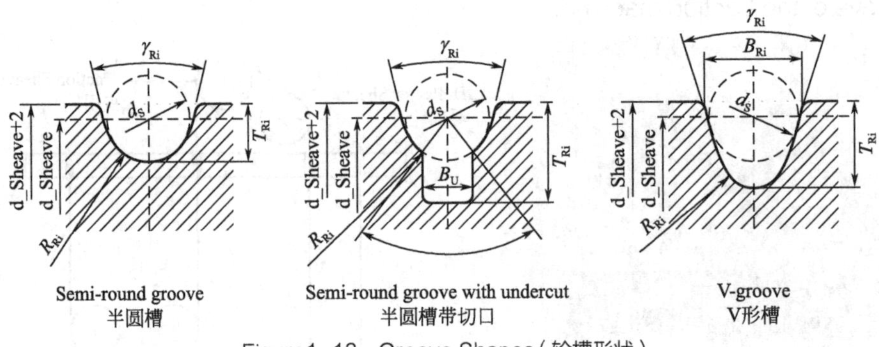

Figure1-13 Groove Shapes（轮槽形状）

Notes and Expressions

1. versatile [ˈvə:sətail] *adj.* 多才多艺的；通用的
2. pulley [ˈpuli] *n.* 滑轮；皮带轮 *vt.* 用滑轮升起
3. circumference [səˈkʌmfərəns] *n.* 圆周；周长
4. gear train 齿轮系；齿轮组
5. loop [lu:p] *vi.* 打环；翻筋斗；循环；回路 *n.* 环；圈
6. counterweight [ˈkauntə,weit] *n.* 平衡物；对重装置；配重

7. potential energy 势能；位能
8. see-saw ['si:sɔ:] *n.* 跷跷板 *adj.* 摇摆不定的
9. solenoid ['səulənɔid] *n.* 螺线管；筒形线圈；电磁线圈
10. asbestos [æz'bestɔs] *n.* 石棉 *adj.* 石棉的
11. redundancy [ri'dʌndənsi] *n.* 过多；多余；冗长，累赘；裁员；
12. helical ['helikəl] *adj.* 螺旋形的
13. layman ['leimən] *n.* 外行人；门外汉；
14. worm [wə:m] *n.* 蠕虫；蜗杆 *vt.* 使蠕动 *vi.* 慢慢前进
15. more often than not 通常，多半

Section 3 Weight Balance System 重量平衡系统

■ Counterweight 对重

The first improvement that appeared in cable-lifted elevators was the counterweight. Lifting the elevator car by itself requires a considerable amount of work because the car's gravitational potential energy increases as it rises. It would be nice to get back this stored energy when the car descends and it's possible to use that energy to lift a counterweight.

The counterweight in an elevator descends when the car rises and rises when the car descends. Because the two objects have similar masses, the total amount of mass that is rising or falling as the elevator moves is almost zero. The overall gravitational potential energy of the elevator is not changing very much; it's simply moving around between the various parts of the machine. The counterweight balances the car so that it takes very little power to move the system. The elevator car and counterweight resemble a balanced seesaw, which requires only a tiny push to make it move.

The counterweight (as shown in Figure 1-14) on most elevators hangs from its own cable attached to the elevator car. That cable travels from the car, over pulleys at the top of the elevator shaft, and down to the counterweight. The counterweight is usually equal to the mass of the empty elevator car plus about 40% of the elevator's rated load. Thus, when the elevator is 40% filled, the counterweight will exactly balance the car and very little work will be done in raising or lowering the car.

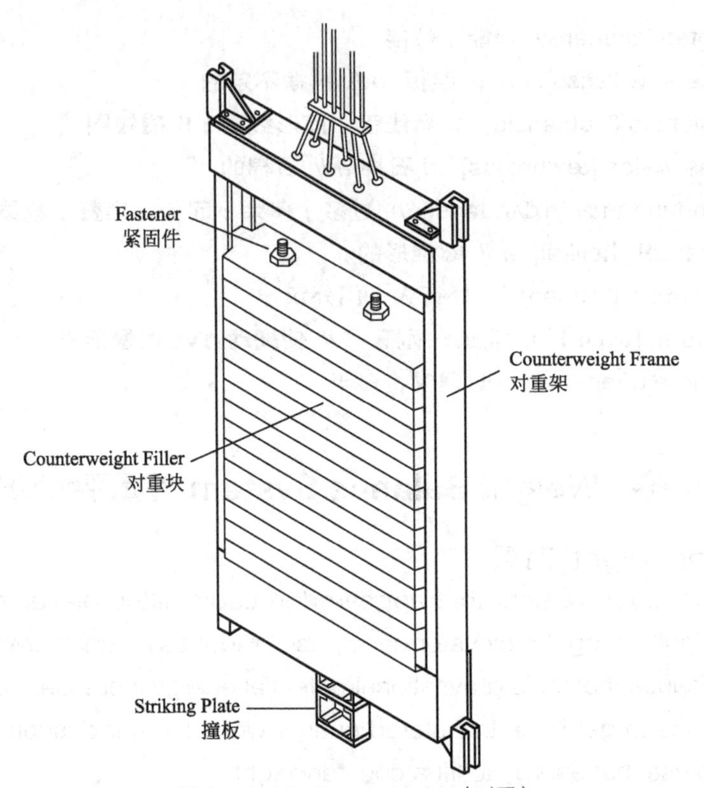

Figure1-14 Counterweight（对重）

▪ Weight Compensation Device 重量补偿装置

Elevators with more than 100ft (30 m) of travel should have a weight compensation system. This is a separate set of cables or a chain attached to the bottom of the counterweight and the bottom of the elevator cab. This makes it easier to control the elevator, as it compensates for the differing weight of cable between the counterweight side and the car side caused by the steel wire rope.

If the elevator cab is at the top of the hoist-way, there is a short length of hoist cable above the car and a long length of compensating cable below the car and vice versa for the counterweight. If the compensation system uses cables, there will be an additional sheave in the pit below the elevator, to guide the cables. If the compensation system uses chains, the chain is guided by a bar mounted between the counterweight railway lines.

Notes and Expressions

1. ASME A17.1—2000 is Safety Code for Elevators and Escalators drawn by American Society of Mechanical Engineers/ 627 pages / ISBN: 0791826325, and has been replaced by ANSI/ASME A17.1—2008 now.

2. 1 ft =12in=12 × 25.4mm=305mm

3. compensation [ˌkɔmpen'seiʃən] *n.* 补偿；报酬；赔偿金

4. gravitational potential energy 重力势能

5. mass [mæs] *n.* 块，团；群众；质量

【Self Evaluation】/【自我评价】

1. In general, hydraulic elevators are suitable for low-rise buildings whereas, the traction elevators are best suited to higher buildings. (　)

(A) True　(B) False　(C) Maybe　(D) Not always

2. The brake on the driving machine____________.

(A) serves to stop the elevator car.

(B) only sets when the mainline switch is opened.

(C) is applied by spring force when electric power is removed.

(D) is attached to the crosshead.

3. In this chapter, the counterweight weighs about the same as the car filled to______percent capacity.

(A) 40　(B) 45　(C) 50　(D) 55

4. The most common elevator brake is made up of ____________, brake shoes with linings, and a solenoid assembly.

(A) isolation pads　(B) a compressive spring assembly

(C) drive motor　(D) diverter pulley

5. The greater the number of wires in a strand the less flexibility in the wire rope. (　)

(A) True　(B) False　(C) Maybe　(D) Not always

6. The ________ balances the car so that it takes very little power to move the system.

(A) wire rope　(B) brake　(C) guide rail　(D) counterweight

7. According to their function, elevator is made up of eight systems:

______________，______________，______________，______________、

______________，______________，______________，______________。

【Extensive Reading】/【拓展阅读】

History of Elevators
电梯历程

▪ The Ancient Age 远古时代

Since the dawn of time, humans sought the way for more efficient vertical transportation of freight and passengers to different levels. These devices for transport goods up and down represent first elevators.

The elevator is not exactly a new idea. Its pioneer form may be traced back to the Middle Ages, as early as the 3rd century BC, when heavy weights were lifted by aid of an apparatus called teagle, usually worked by hand power, animal power or sometimes water-driven mechanisms.

In order to ascend more easily, man devised the stairway and ladder, from which, in turn, was developed the escalator and elevator, in order to further eliminate physical effort.

■ 1st Industrial Revolution 第一次工业革命（蒸汽时代）

Modern elevators were developed during the 1800s. These crude elevators slowly evolved from steam driven to hydraulic power. They were used for conveying materials in factories, warehouses and mines. Hydraulic elevators were often used in European factories.

In 1823, two architects Burton and Hormer built an "ascending room" as they called it, this crude elevator was used to lift paying tourists to a platform for a panorama view of London.

In 1835, architects Frost and Stutt built the "Teagle", a belt-driven, counter-weighted, and steam-driven lift was developed in England.

In 1846, Sir William Armstrong introduced the hydraulic crane, and in the early 1870s, hydraulic machines began to replace the steam-powered elevator. The hydraulic elevator is supported by a heavy piston, moving in a cylinder, and operated by the water (or oil) pressure produced by pumps.

In 1852, Elisha Graves Otis introduced the first safety contrivance for elevators. Otis established a company for manufacturing elevators and went on to dominate the elevator industry. Today the Otis Elevator Factory is the world's largest manufacturer of vertical transport systems.

On March 23, 1857 the first Otis passenger elevator was installed at 488 Broadway in New York City.

The development of practical wire rope by John Roebling had a strong influence on elevator design. The year 1861 marked the beginning of using multiple ropes for suspension. Each rope was capable of supporting the full weight of the elevator. This practice is still in use today. The only known documented instance of all the ropes breaking at once and allowing the elevator to free fall in modern times occurred when an airplane ran into the Empire State building in 1945.

- 2nd Industrial Revolution 第二次工业革命（电气时代）

Revolution in elevator technology began with the invention of hydraulic and electricity. Motor technology and control methods evolved rapidly and electricity quickly became the accepted source of power. The safety and speed of these elevators were significantly enhanced. The first electric elevator was built by the German inventor Wener Von Siemens in 1880.

In 1887, an electric elevator was developed in Baltimore, using a revolving drum to wind the hoisting rope, but these drums could not practically be made large enough to store the long hoisting ropes that would be required by skyscrapers.

Motor technology and control methods evolved rapidly. In 1889 came the direct-connected geared electric elevator, allowing for the building of significantly taller structures. By 1903, this design had evolved into the gearless traction electric elevator, allowing hundred-plus story buildings to become possible and forever changing the urban landscape.

- 3rd Industrial Revolution 第三次工业革命（信息时代）

Many changes in elevator design and installation was made by the great advances in electronic systems during World War II.

The first fully automatic elevators were installed in the mid 1950s and very soon became the standard.

Since 1955, the Safety Code for Elevators and Escalators has prohibited use of winding drum machines except for freight elevators with less than 40 feet of rise and speed of 50 ft/min. The traction elevator; also replace the water rope hydraulic elevators, for medium-rise installation.

- As Good As It Gets 尽善尽美

Today, there are intricate governors and switching schemes to carefully control cab speeds in any situation. "Buttons" have been giving way to keypads. Virtually all commercial elevators operate automatically and the computer age has brought the microchip-based capability to operate vast banks of elevators with precise scheduling, maximized efficiency and extreme safety. Elevators have become a medium of architectural expression as compelling as the buildings in which they're installed, and new technologies and designs regularly allow the human spirit to soar!

Notes and Expression

1. teagle ['ti:gl] *n.* 卷扬机，绞盘机；绞辘

Chapter 2

Structure of Elevator Ⅱ
电梯结构Ⅱ

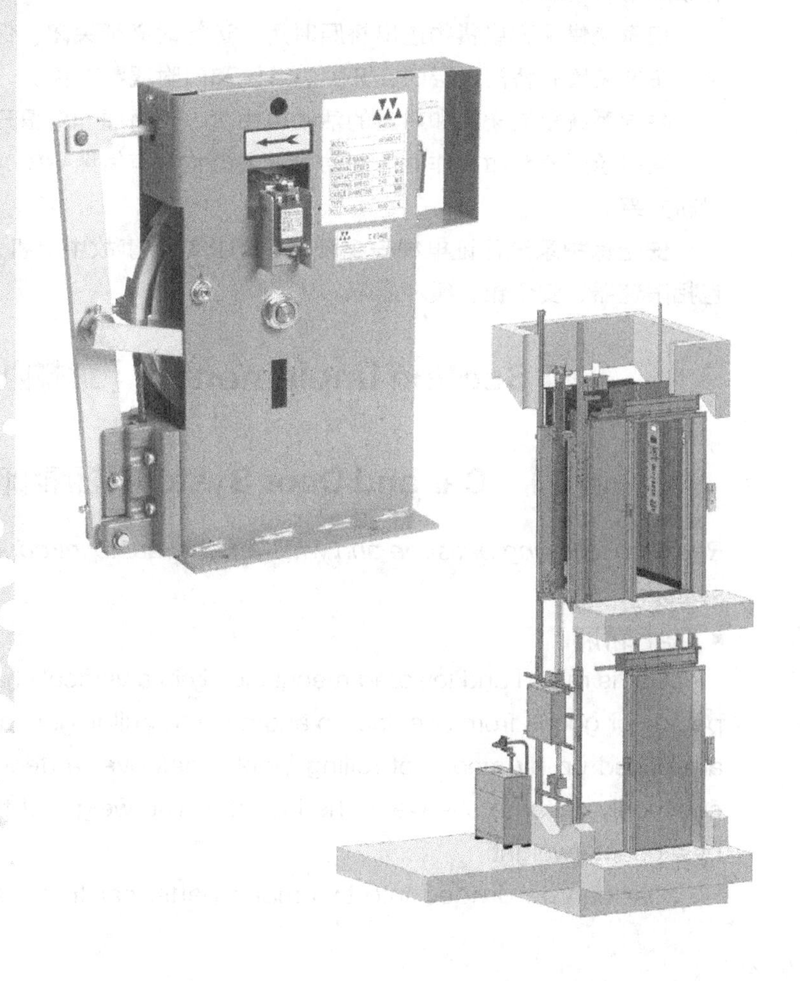

【Content Description】/【内容描述】

电梯是机电技术高度结合特种设备，了解电梯的结构是制造、安装、使用和维护的前提。本章继续第1章的内容，讲述电梯的结构和原理，主要内容为八大系统中的轿厢系统，门系统，导向系统，安全保护系统和电气控制系统。

本章内容主要涉及电梯主要部件的功能和相互关系，这些知识对将来作为电梯技术人员进行交流和阅读英语文献时至关重要。通过本章的学习，应掌握轿厢、层门和轿门、导轨及导轨支架、导靴、限速器、安全钳、缓冲器的英语词汇，能阅读并翻译常用的描述句型。

【Related Knowledge】/【知识准备】

轿厢系统用以装运并保护乘客或货物的组件，是电梯的工作部分，主要构件有轿厢架和轿厢体。

门系统供乘客或货物进出轿厢时用，运行时必须关闭，保护乘客和货物的安全，主要装置有轿门、层门、开关门系统和门附属零部件。

导向系统限制轿厢和对重的活动自由度，使轿厢和对重只能沿着导轨上下运动，同时承受安全钳工作时的制动力，主要构件有轿厢导轨、对重导轨、导靴和导轨支架。

安全保护系统保证电梯安全使用，防止意外事故中危机人身和设备的安全，包括限速器、安全钳、缓冲器等。

【Section Implement】/【章节内容】

Section 1 Car and Door System 轿厢和门系统

Read the following passage and write an abstract to introduce traction elevator systems.

▪ Car 轿厢

Car is raised and lowered mechanically in a vertical shaft in order to move people or goods from one floor to another in a building. In a "traction" lift, cars are pulled up by means of rolling steel ropes over a deeply grooved pulley, commonly called a sheave in the industry. The weight of the car is balanced by a counterweight.

Car can be divided into two major parts: car frame and car body. Car body (as shown in Figure 2–1) is composed of car floor, car walls and car roof. The supporting frame (as shown in Figure 2–2) of an elevator car to which are attached the car platform, guide shoes, elevator car safety, hoisting ropes or sheaves, and/or associated equipment.

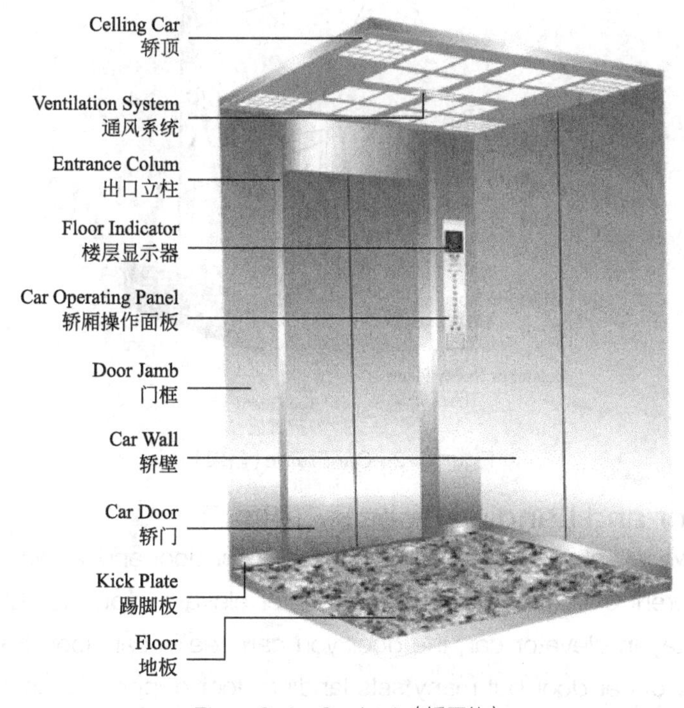

Figure2–1 Car body（轿厢体）

The Crosshead is the upper member of the car frame. The Stiles are the vertical member of the car sling, one on each side, which fastens the crosshead to the safety plank.

The Brace Rod is a rod extending from the elevator platform framing to another part of the elevator car frame or sling for the purpose of supporting the platform or holding it securely in position. Brace rods are supports for the outer corners of the platform, each of which tie to upper portions of the stile.

The platform isolation is rubber or other vibration absorbing material which reduces the transmission of vibration and noise to the platform. These pads are often replaced when modernizing as new isolation is more resilient and helps to reduce vibration and improve the comfort of the ride for passengers.

The bolster is the bottom horizontal member of a hydraulic car sling, to which the platen plate attaches. The safety plank for a traction elevator, bottom member of a sling, contains the safety.

Figure2-2 Car Frame（轿架）

■ Car Door and Landing Door 轿门和层门

When we talk about elevator door, there are car door and landing door which are two different conceptions. Before you enter elevator, you face landing door; when you stay in elevator car, the door you can see is car door. Each elevator has one pair of car door but many sets landing door depending on the height of building. Car door is active door while landing door is passive door.

Landing door protect riders from falling into the shaft. Figure2-3 depicts the four major types of landing doors used in elevator. The passenger types are the center-opening, the single slide and the two-speed sliding door (two panels overlapped when open). The typical car landing door is the vertical bi-parting, the top half rises and the lower half depresses at the same time to form an opening almost as wide as the car itself.

The general descriptions of the various forms of protective devices for elevator door can be broken down into two categories. The earliest type and still very popular is the mechanical protective edge, usually consisting of a strip at the front edge of the car door (see figure2-4). If the strip is moved back during the closing operation, a switch is actuated, causing the doors to stop and reverse. Materials used for the edges range from soft rubber to stiff plastic and, in some applications, metal. As the edge may be operated many hundreds of times during a day, subject to many kinds of contact and abuse, the entire mechanism must be "rugged."

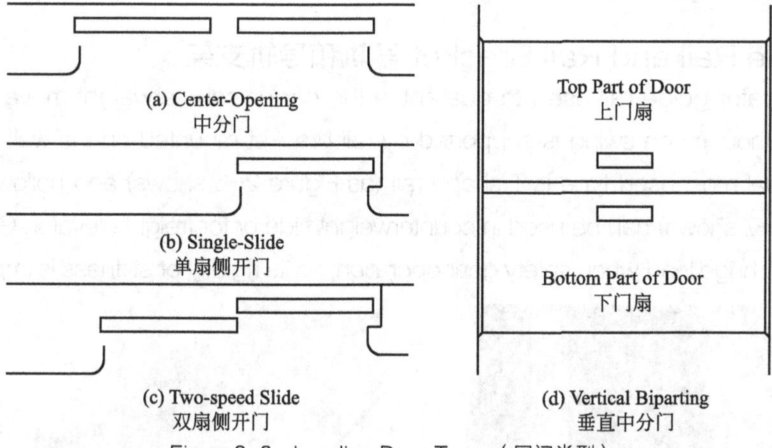

Figure2-3 Landing Door Types (层门类型)

Another type is infrared protective system, a kind of contactless method (see figure2-5). New door safety systems use infrared beams that cover the doorway of the elevator. When the beams are broken, the doors retract. Unlike old mechanical systems, the new doors do not come into contact with either passengers or objects.

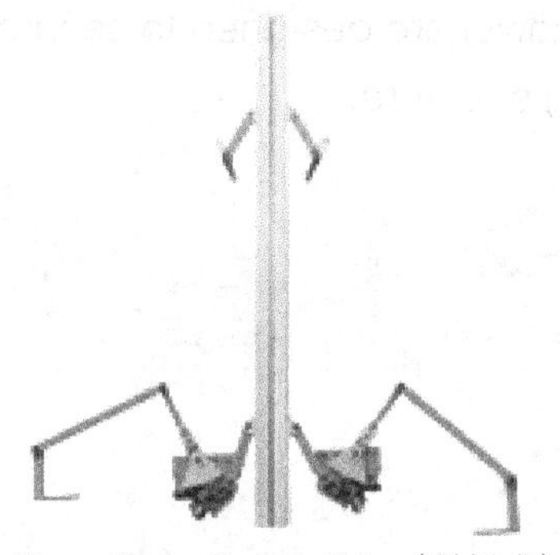

Figure2-4 Safety Edge (触板式)

Figure2-5 Infrared Type (红外式)

Notes and Expressions

1. resilient [ri'ziliənt] *adj.* 有弹力的
2. stile [stail] *n.* 门梃；立梁；侧板；立柱
3. bi-parting 上下对开式；垂直中分式
4. category ['kætigəri] *n.* 种类，分类；范畴
5. infrared [ˌinfrə'red] *n.* 红外线 *adj.* 红外线的
6. active / passive 主动的/移动的（从动的）

Section 2 Guide System 导向系统

▪ Guide Rail and Rail Bracket 导轨和导轨支架

Elevator guide rail used to guarantee the car or counterweight move up and down without much swing is supported by rail bracket mounted on the wall of shaft. The actual most used type is T shape rail (as Figure 2-6 shows) and hollow rail (as Figure 2-7 shows) can be used in counterweight side or for freight elevator. Guide rail will suffer huge load when safety gear operation, so its rigidity or stiffness is important.

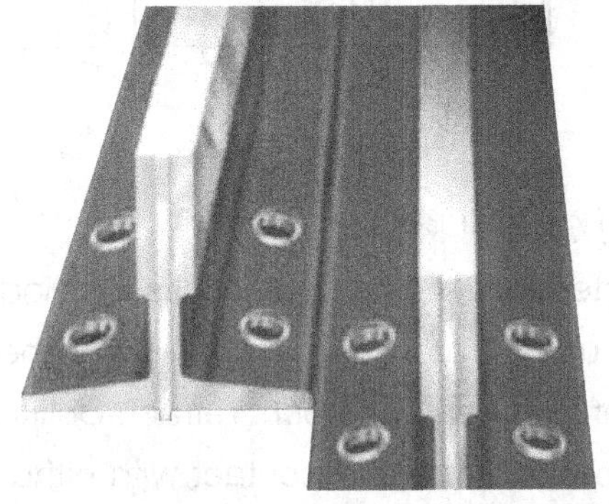
Figure2-6 T Shape Rails (T形导轨)

Figure2-7 Hollow Rails (空心导轨)

Guide rail brackets (as Figure 2-8 shows) are designed to securely attach car and counterweight rails to the building structure.

Figure2-8 Rail Bracket (导轨支架)

▪ Guide Shoe 导靴

The guiding shoes used to guide the car and counterweight on the guide rails may be either sliding or roller type. Sliding guide shoes are often used for rated speed up to 150 ft/min (about 0.76m/s). In some cases, heavy duty freight elevators may use sliding guide shoe for higher speeds. Example of sliding guide shoe is shown in Figure2-9.

Roller guide shoes are usually used for rated speeds over 150 ft/min but may be used for any speed. They are sometime used for slow speed to avoid the need to lubricate the guide rails. Roller guide shoe is shown in Figure2-10.

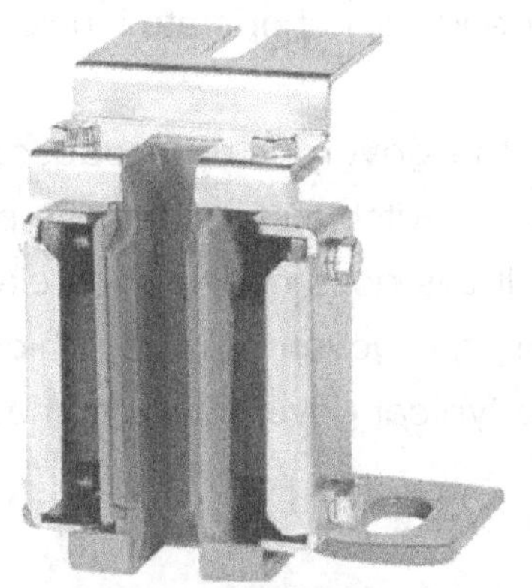

Figure2-9　Sliding Guide Shoe
（滑动导靴）

Figure 2-10　Roller Guide Shoe
（滚动导靴）

If guide shoes are not maintained or adjusted properly your elevator will tend to rock, sway, squeak or even rumble its way through the hoistway.

Section ❸ Safety Protection System 安全保护系统

Elevators are used by thousands of people every day. Responsible building owners all over the world are doing their part to keep people safe by upgrading their existing vertical transportation equipment to meet all local and national safety codes.

In the world of Hollywood action movies, hoist ropes are never far from snapping in two, sending the car and its passengers hurdling down the shaft. In actuality, there is very little chance of this happening. Elevators are built with several redundant safety systems that keep them in position.

The first line of defense is the rope system itself. Each elevator rope is made from several lengths of steel material wound around one another. With this sturdy structure, one rope can support the weight of the elevator car and the counterweight on its own. But elevators are built with multiple ropes, typically between four and eight. In the unlikely event that one of the ropes snaps, the rest will hold the elevator up.

Even if all of the ropes were to break, or the sheave system was to release them, it is unlikely that an elevator car would fall to the bottom of the shaft. Roped elevator cars have built-in braking systems, or safeties, that grab onto the rail when the car moves too fast.

▪ Overspeed Governor 限速器

The governor safety system is an independent safety system that is not intended to regulate speed or stop the elevator during normal operation. The movement of the car passes to the governor and will not operate it unless the car overspeeds.

As the car or counterweight moves, the governor rope causes the governor to spin. At a set speed, it will open a switch that will remove power from the driving machine motor and brake. If this does not stop the elevator going down, the governor will trip, gripping the governor rope, which will set the safety and stop the elevator. Some typical governors are shown in Figure2-11.

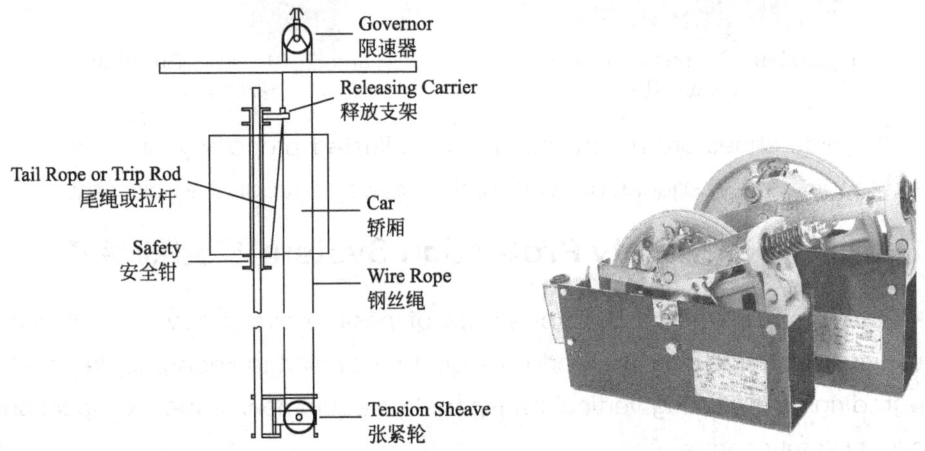

Figure2-11　Governor and it's Location（限速器及其位置）

▪ Safety Gear 安全钳

Safety gear is braking system attached on the elevator car, running up and down the elevator shaft and grabs onto the rails when emergency. Some safeties clamp the rails, while others drive a wedge into notches in the rails. Typically, safeties are activated by a mechanical speed governor.

The safety gear housing is usually mounted on the frame using transverse bars or the like. To prevent accidental gripping of the wedges when the elevator car moves laterally, the safety gear housing is allowed to move laterally along the bars.

The safety gear is provided with guiding means placed on either side of the guide rail so that the clearance between them and the guide rail is smaller than the clearance between the wedge and the guide rail. Thus, when the lateral displacements of the frame of the elevator car relative to the guide rail exceeds the distance between the guiding means and the guide rail, the safety gear housing is moved laterally and a minimum distance between the wedge and the guide rail is maintained.

The safety gear can be divided into two types: the instantaneous type and the progressive type (as Figure 2-12 shows), the former for lower rated speed elevator and the latter for higher rated speed elevator.

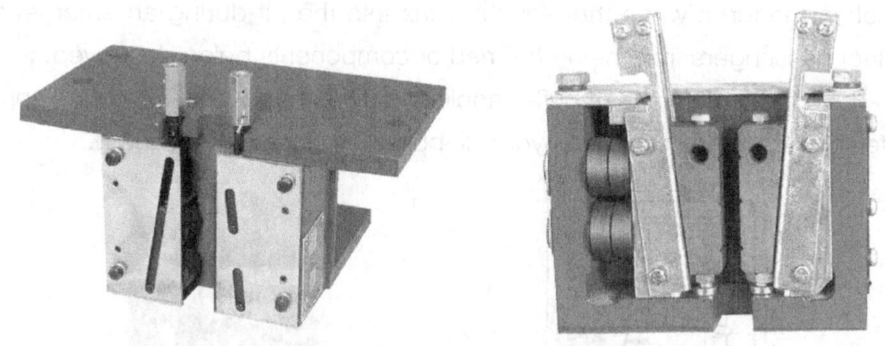

Figure2-12 Safety Gear(安全钳)

▪ Rope Gripper夹绳器

The primary overspeed governor and safety can protect the car from falling to the pit, but it will not stop the ascending car. The A17.1-2000 Code will require ascending car protection and unintended motion protection. This may be accomplished by using a separate brake on the rope or a means of applying a brake on the drive sheave for old elevator. In order to stop the ascending car, we now have another device; called rope gripper, see figure2-13.

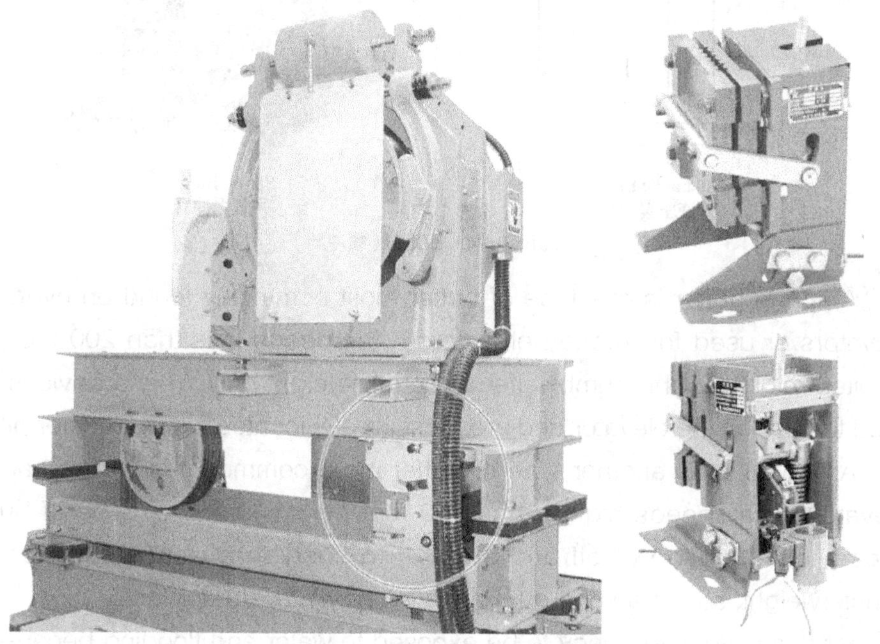

Figure2-13 Rope Gripper(夹绳器)

■ Buffer 缓冲器

Buffer is an important part of elevator safety system, and it is the last barrier if elevator out of control. Buffers can reduce the shock to a great extent or soften the force when the elevator runs into the pit during an emergency, protect passengers from being harmed or components being destroyed.

There are three types of buffer applied in different elevator, which are spring buffer, polyurethane buffer and hydraulic buffer (as Figure 2-14 shows).

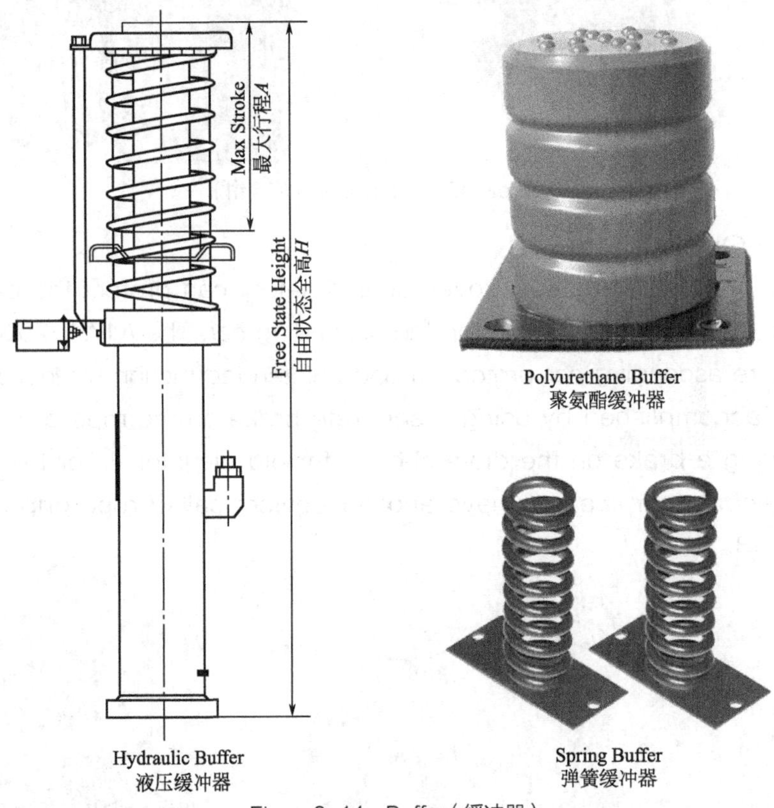

Figure2-14 Buffer(缓冲器)

A spring buffer is one type of buffer most commonly found on hydraulic elevators or used for traction elevators with speeds less than 200 feet per minute, freight weight, dumbwaiter or counterweight only. These devices are used to cushion the elevator and are most always located in the elevator pit.

An oil buffer is another type of buffer more commonly found on traction elevators with speeds higher than 200 feet per minute. This type of buffer uses a combination of oil and springs to cushion a descending car or counterweight, so it is an energy dissipation type elevator buffer.

Buffers have a tendency to be exposed to water and flooding because of their location in the pit. They require routine cleaning and painting to assure they maintain their proper performance specifications. Oil buffers also need oil level check and change if exposed to flooding.

Notes and Expressions

1. transverse [træns'və:s] *adj.* 横向的；横断的 *n.* 横断面
2. lateral ['lætərəl] *adj.* 侧面的；旁边的；横向的；横斜的
3. grab [græb] *vt.&vi.* 攫取，抓取；捕获，逮住
4. grip [grip] *n.* 紧握；柄；支配 *vt.* 紧握；夹紧 *vi.* 抓住
5. polyurethane [ˌpɔli'jurəˌθein] *n.* 聚氨基甲酸乙酯；聚氨酯
6. cushion ['kuʃən] *n.* 垫子；胶垫 *vt.* 缓和……的冲击
7. dissipation [disi'peiʃən] *n.* 驱散；消散；挥霍，浪费；耗散

Section 4 Electrical Control System 电气控制系统

Automatic elevators began to appear as early as the 1930s, their development being hastened by skyscrapers increase in large cities such as New York and Chicago. These electromechanical systems used relay circuits to control the speed, position and door operation of an elevator. Relay-controlled elevator systems remained common until the 1980s, and their gradual replacement with microprocessor based controls which are now the industry standard.

▪ Car Operation Panel 轿厢操纵面板

The car operating panel (as shown in Figure 2-15) is equipped with all the components necessary for the operation of an elevator. With the different button systems an enormous range of design is available. The faceplate is normally made of satin stainless steel, or hairline finished stainless steel if needed.

The double banked car operating panel is advisable for a higher number of floors, because the single banked version would be too long. As the length of the faceplate with a double banked panel is smaller than a single banked panel, the buttons are accessible by all.

▪ Control Cabinet 控制柜

Elevator control cabinet (as Figure 2-16 shows) is important to elevator and can be seen as brain of people, receiving signals from COP & calling board and giving instructions to traction machine all the time.

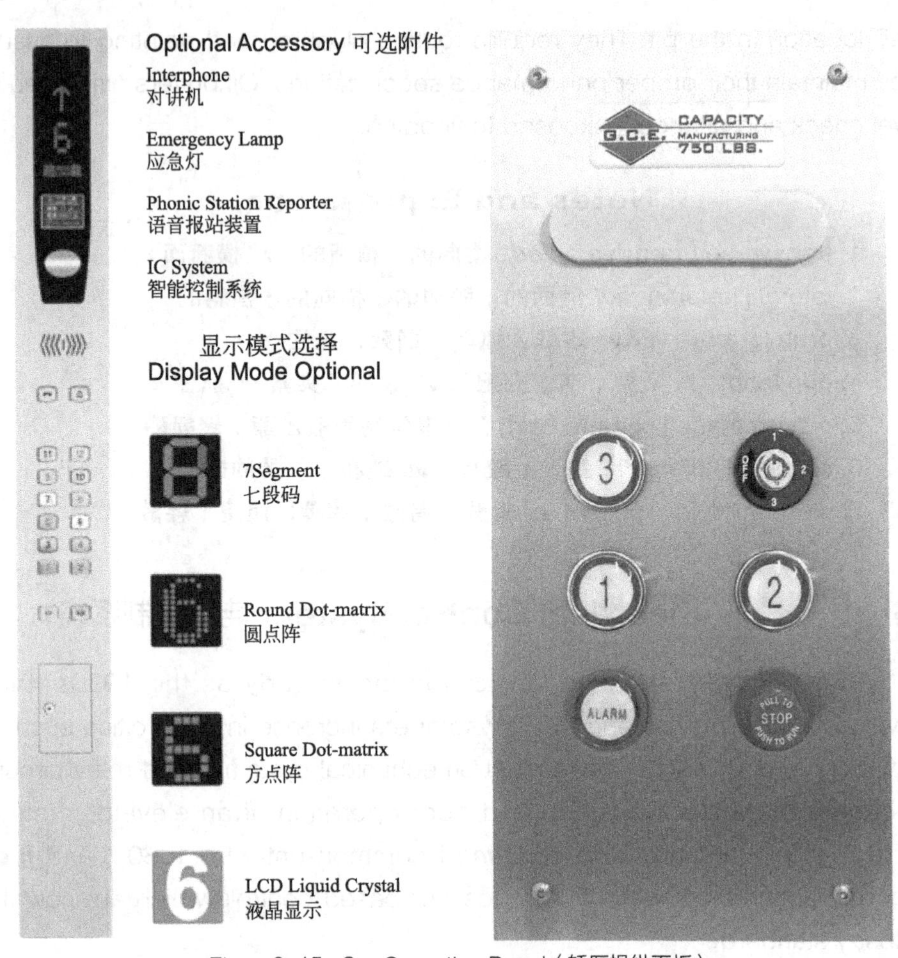

Figure2-15 Car Operation Panel（轿厢操纵面板）

Elevator control cabinet locates in machine room for most elevators. It will be mounted in shaft for machine-room-less elevator.

There appears elevator with "destination control system" function nowadays, a destination operating panel where a passenger registers their floor calls before entering the car. Although travel time is reduced passenger waiting times may be longer.

- **Calling Board 召唤盒**

Calling board (as Figure 2-17 shows) or Hall calling board can be found in each landing. Most landing calling boards have at least two calling buttons for upward and downward except the terminal station. The structure of calling board is very similar with COP.

Figure2-16　Control Cabinet
（控制柜）

Figure2-17　Calling Board
（召唤盒）

Notes and Expressions

1. COP（Car operation panel）轿厢操纵面板
2. skyscraper ['skaiˌskreipə]　*n.* 摩天楼，超高层大楼
3. hairline finished stainless steel　发纹处理不锈钢
4. satin stainless steel　镜面不锈钢

【Self Evaluation】/【自我评价】

1. Oil buffers are required on __________.
(A) the car only.
(B) the counterweight only.
(C) both car and counterweight.
(D) both car and counterweight for rated speed over 200 ft/min.
2. The purpose of the overspeed governor is to ___________.
(A) control the speed of the elevator.
(B) cause the power to be removed and the safety to set when overspeed

occurs.

(C) assure that the elevator is operating as fast as possible.

(D) signal the controller to cause the elevator to move.

3. Standing inside the elevator, what can you see? (   )

(A) Landing door.

(B) Car door.

(C) Both car door and landing door.

(D) Neither car door nor landing door.

4. A retractable safety edge on a horizontal sliding door ____________.

(A) Requires contact to cause the door to stop and reopen.

(B) Does not require contact to cause the door to stop and reopen.

(C) May or may not require contact to cause the door to stop and reopen.

(D) Is used on most new elevators.

5. The Crosshead is the _________ member of the car frame.

(A) upper (B) below

(C) middle (D) side

6. Car can be divided into two major parts: __________ and ___________.

7. The infrared protective system of car door is a kind of _________method.

8. The guiding shoes used to guide the car may be either sliding or _________type.

9. The safety gear can be divided into two types: the instantaneous type and the progressive type, ___________________for lower rated speed elevator.

【Extensive Reading】/【拓展阅读】

Elevator Company
电梯公司

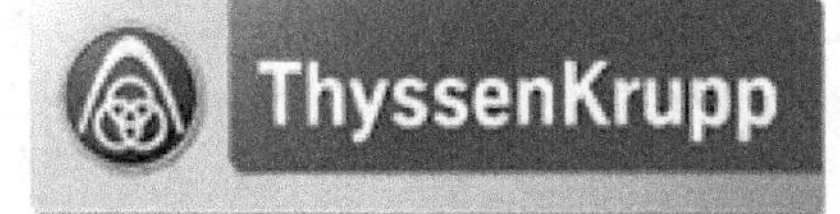

ThyssenKrupp Elevator
蒂森电梯

ThyssenKrupp Elevator is one of the world's leading elevator companies. With sales of approximately5.2 billion and 44000 employees at 900 locations, ThyssenKrupp Elevator's aim is to continue to grow strategically and profitably. Innovation and quality are hallmarks of our products and services and assure lasting customer relationships.

Mitsubishi Elevator
三菱电梯

A major name in elevators and escalators since the 1930s, Mitsubishi Electric has built a reputation for creating breakthroughs that make getting around more comfortable, safe, and even inspiring. They're also a leading manufacturer of moving walks and home elevators.

OTIS Elevator
奥蒂斯电梯

Otis Elevator Company is the world's leading manufacturer, installer and maintainer of elevators, escalators and moving walkways—a constant, reliable name for more than 150 years. Otis elevators and escalators touch the lives of people in more than 200 countries around the world.

Schindler Elevator
迅达电梯

Established by Mr. Robert Schindler in 1874, Schindler—the Elevator and Escalator Company is headquartered in the picturesque town of Lucerne, Switzerland. Schindler is the world's leading manufacturer of escalators and the second largest supplier of elevators.

Dedicated to People Flow 

KONE Elevator
通力

KONE is one of the global leaders in the elevator and escalator industry. The company has been committed to understanding the needs of its customers for the past century, providing industry-leading elevators, escalators and automatic building doors as well as innovative solutions for modernization and maintenance.

SIGMA Elevator
星玛电梯

Starting in 1968, "Goldstar" elevators, and later on, "LG" elevators, were counted among the world's top elevator brands.

TOSHIBA
Leading Innovation >>>

TOSHIBA Elevator
东芝电梯

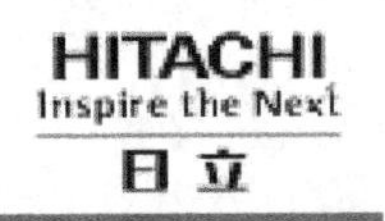

Hitachi Elevator
日立电梯

Edunburgh Elevator
爱登堡电梯

Bester Elevator
百思特电梯

CANNY Elevator
康力电梯

CANNY Elevator Co.,Ltd is a modernized and professional elevator manufacturer which integrates the design, development, manufacturing, sales, installation, and maintenance into a whole. On 12th Mar., 2010, Canny Elevator Co., Ltd. came into Shenzhen Stock Exchange A share market successfully.

DNDT Elevator
东南电梯

Dongnan Elevator Co.,Ltd, abbreviated as DNDT, founded in 1998 and located in SIP (Sino-Singapore Industrial Park).

IFE Elevator
快意电梯

IFE Elevator was established and approved by the Ministry of Construction of the PRC, in 1988.

Fulingda Elevator Co., Ltd

Fulingda Elevator
富菱达电梯

Fulingda Elevator Co., Ltd was founded in 1980s. Located in Panyu district in Guangzhou city, it is subordinated to Zhujiang Industry Group Co., Ltd.

GUANGRI

Guangri Elevator
广日电梯

Guangzhou Guangri Elevator Industry Co., Ltd., a key pillar enterprise under Guangzhou Guangri Group Co., Ltd., is one of the largest bases of production of complete sets of elevators in South of China.

三洋电梯（珠海）有限公司
SANYO ELEVATOR (ZHUHAI)CO.,LTD.

Sanyo Elevator
三洋电梯

As a foreign-owned enterprise, Sanyo Elevator (Zhuhai) Co., Ltd. was established in 2006. Wit is specialized in one step service of elevator R & D, manufacture, sales, installation, maintenance, and so on.

SJEC 江南嘉捷

SJEC Elevator
嘉捷电梯

SJEC Corporation (SJEC) was set up in 1992 in Suzhou Industrial Park (SIP), Suzhou. China specialized in sales, design, and manufacturing and installation, maintenance and service of state-of-art of elevators, escalators, passenger conveyors and Car parking system.

CNYD·BLT
沈阳博林特电梯股份有限公司
SHENYANG BRILLIANT ELEVATOR CO.,LTD.

BLT Elevator
博林特电梯

As one of the core industry of Yuanda Enterprise Group, BLT elevator has its own brand and intellectual property. After striving few years, BLT has been involved into the leading team of global elevator industry.

Shenlong Elevator
申龙电梯

Winone Elevator
菱王电梯

Guangdong WINONE Elevator Co.,Ltd. is located in Foshan City, Guangdong Province in 2002.

OMS Elevator
奥美森电梯

Chapter 3

Basic Knowledge of Metal
金属基本知识

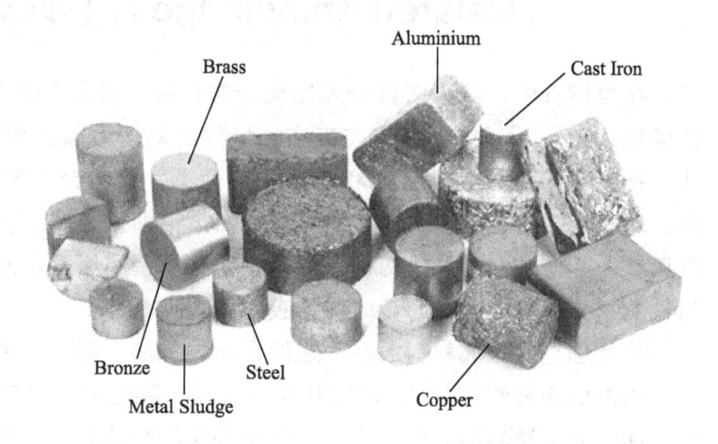

【Content Description】/【内容描述】

金属材料知识是机械加工和制造的基础。而热处理是金属加工的主要方式，对金属特性有重要影响，尤其是硬度。从原材料到零部件，根据需要采用不同的加工方式及工具，包含种类众多的机床。本章通过以上内容，训练大家对专业英语的认知。

本章共三节，第一节是金属分类，训练学生的理解能力，抓住句子主要意思的能力。第二节为金属特性与热处理，训练学生的词组意识，将一组单词看做一个整体不但节省时间，还利于理解。同时要训练学生猜测词义的能力，在有新单词时，尽量不查字典猜测句子意思的能力。第三节介绍金属加工机床，在本节中学生要注意训练自己的阅读习惯，避免反复回看等不良习惯，养成从头到尾不间断阅读的习惯。

【Related Knowledge】/【知识准备】

在自然界中，绝大多数金属以化合态存在（多数是氧化物及硫化物），少数金属例如金、铂、银以游离态存在。世界上对于金属元素的分类基本有两种方法；把金属元素分为铁金属和非铁金属两类，或者分为黑色金属和有色金属两类。对于众多的有色金属，又将其分为四类，即重金属、轻金属、贵金属和稀有金属。合金，是由两种或两种以上的金属与非金属经一定方法所合成的具有金属特性的物质。一般通过熔合成均匀液体并凝固而得。

热处理是将金属材料放在一定的介质内加热、保温、冷却，通过改变材料表面或内部的金相组织结构，来控制其性能的一种金属热加工工艺。金属热处理工艺大体可分为整体热处理、表面热处理和化学热处理三大类。钢铁整体热处理大致有退火、正火、淬火和回火四种基本工艺。

机械加工主要有手动加工和数控加工两大类。手动加工是指通过机械工人手工操作铣床、车床、钻床和锯床等机械设备来实现对各种材料进行加工的方法。手动加工适合进行小批量、简单的零件生产。 数控加工是指机械工人运用数控设备来进行加工，这些数控设备包括加工中心、车铣中心、电火花线切割设备、螺纹切削机等。目前，绝大多数的机加工车间都采用数控加工技术。数控加工以连续的方式来加工工件，适合于大批量、形状复杂的零件。

【Section Implement】/【章节内容】

Section ❶ Classification of Metals 金属分类

Read the passage and write an abstract to describe the classification of metal.

In industry today, there are more than a thousand different metals being used to manufacture products. The modern automobile has more than one hundred different metals used in its construction. An attempt will be made in this passage to give an understanding of the basic classification of metals.

Metals were formerly thought to be those elements that had a metallic luster and were good conductors of heat and electricity. Actually, metals are generally defined as those elements whose hydroxides form bases (such as sodium or potassium) , and nonmetals' hydroxides form acids (such as sulphur). Metals may exist as pure elements. When two or more metallic elements are combined, they form a mixture called an alloy.

The term alloy is used to identify any metallic system. In metallurgy it is a substance, with metallic properties, that is composed of two or more elements, intimately mixed. Of these elements, one must be a metal. Plain carbon steel, in the sense, is basically an alloy of iron and carbon. Other elements are present in the form of impurities. However, for commercial purposes, plain carbon steel is not classified as an alloy steel.

Although pure metals solidify at a constant temperature, alloys do not. Alloys may be further classified as ferrous and nonferrous. Ferrous alloys contain iron. Nonferrous alloys do not contain iron.

All commercial varieties of iron and steel are alloys. The ordinary steels are thought of as iron-carbon alloys. However, practically all contain silicon and manganese as well. In addition, there are thousands of recognized alloy steels. Examples are special tool steels, steels for castings, forgings, and rolled shapes. The base metal for all these is iron.

Steels are often called by the principal alloying element present. Examples are silicon steel, manganese steel, nickel steel, and tungsten steel. Even nonferrous alloys may contain iron in a small amount, as impurities. Some of the nonferrous alloys are bronze, and brass.

Nonferrous metals are seldom found in the pure state. They must be separated from the gangue before the ore can be reduced. Thus, a process known as ore-dressing is performed. Metals and metal compounds are heavier than the gangue. They settle to the bottom if such a mixture has been agitated in water. This process is similar to the method used by the early miners who panned for gold. However, refinements have been developed to speed up the accumulation of metal compounds by using this principle.

The reverberatory furnace is the type most often used in the smelting of nonferrous metals. This furnace is constructed of refractory brick with a steel structure on the outside. The charge is placed in the furnace and heated indirectly by the flame. Slag inducers or fluxes are added to the charge to reduce oxidation.

Notes and Expressions

1. sulphur ['sʌlfə] *n.* 硫黄；硫黄色 *vt.* 使硫化；用硫黄处理
2. reverberatory furnace 反射炉
3. hydroxide [hai'drɔksaid] *n.* 氢氧化物
4. alloy [ə'lɔi] *n.* 合金
5. gangue [gæŋ] *n.* 脉石；矸石
6. luster ['lʌstə] *n.* 光泽；光彩
7. sodium ['səudiəm] *n.* 钠
8. potassium [pə'tæsjəm] *n.* 钾
9. Metals were formerly thought to be those elements that had a metallic luster and were good conductors of heat and electricity.
在此句中，metals为主语，that引导的定语从句修饰宾语elements。
10. In metallurgy it is a substance, with metallic properties, that is composed of two or more elements, intimately mixed.
此句中的in metallurgy为状语，with metallic properties为插入语，由that引导的定语从句修饰表语a substance。译文应为：在冶金学中，合金是由两种或两种以上的元素均匀混合、具有金属特性的物质。

Section 2 Properties of Metals and Heat Treatment 金属特性与热处理

Read the passage and describe the relation between heat treatment and property of metal.

Metals have properties that distinguish them from other materials. The most important of these properties is strength, or the ability to support weight without bending or breaking. This property combined with toughness, or ability to bend without breaking, is important. Metals also have advantages regarding resistance to corrosion. They are responsive to heat treatment.

Metals can be cast into many shapes and sizes. They can be welded, hardened, and softened. Metals also possess another important property—recycling and reuse. When a particular product is discarded, it can be cut into convenient sections. These sections can be put into a furnace, remelted, and used in another product.

The properties of metals may be classified in three categories: chemical properties, mechanical properties, and physical properties. Here we will emphasize the primary mechanical properties of metals. In understanding the related areas of metalworking and methods used today, the mechanical properties of metals are of the utmost importance.

The hardness of metals varies greatly. Some, like lead, can be in-dented easily. Others like tungsten carbide, approach diamond hardness. They are of great value as dies for cutting tools of various types. Heat treatment causes changes in the hardness. Annealed tool steel can readily be machined. Often, this is difficult after it has been hardened and tempered. Annealed brass is comparatively soft. When cold-worked the hardness is greatly increased.

Steel's property of hardening when suddenly cooled from a heated condition has been known for many centuries. This field about metal is heat treatment. Heat treatment is a part of metallurgy. It changes the structures of metals by the application of heat.

Hardening: In any heat-treating operation the rate of heating is important. Heat flows from the exterior to the interior of steel at a definite maximum rate. If steel is heated too fast, the outside of the part becomes hotter than the interior. A uniform structure is hard to obtain.

The hardness that can be obtained from a given treatment depends upon the following three factors: quenching rate, carbon content and work-piece size.

Rapid quenching is needed to harden low carbon and medium plain-carbon steels. Water is generally used as a quench for these steels. For high-carbon or alloy steel, oil is used. Its action is not as severe as that of water. Where extreme cooling is desired, brine is used.

The maximum degree of hardness obtainable in steel by direct hardening is determined largely by the carbon content. Steel with low carbon content will not respond greatly to the hardening process. Carbon steels are generally considered as shallow hardening steels. The hardening temperature varies for different steels. The temperature depends upon the carbon content.

The temperature at which steel is usually quenched for hardening is known as the hardening temperature. It is usually 10℃ to 38℃ above the upper critical temperature at which structural change takes place.

Tempering: Hardening makes high-carbon steels and tool steels extremely hard and brittle and not suitable for most uses. By tempering or "drawing" internal stresses developed by the hardening process are relieved. Tempering increased the toughness of the hardened piece. It also seems to make them more plastic, or ductile.

Annealing: Annealing is heating steel slightly above its critical range and cooling very slowly. Annealing relieves internal stresses and strain caused by previous heat treatment, machining, or other cold-working processes. The type of steel governs the temperature to which the steel is heated for the annealing process. The purpose for which annealing is being done also governs the annealing temperature.

Full annealing is used to produce maximum softness in steel. Machinability is improved. Internal stresses are relieved. Process annealing is also called stress relieving. It is used for relieving internal stresses that have occurred during cold-working or machining processes. Spheroidizing is used to produce a special kind of grain structure that is relatively soft and machinable. This processes generally used to improve the machinability in high-carbon steels and in wire-drawing processes.

Normalizing: Normalizing is a process used to relieve the internal stresses due to hot-working, cold-working, and machining. The process consists of heating steel slightly above the upper critical range 30℃ to 50℃ and cooling to room temperature after holding for a while. This process is usually used with low and medium-carbon as well as alloy steels. Normalizing removes all previous effects due to heat treatment.

Notes and Expressions

1. spheroidizing ['sfiərɔidaiziŋ] *n.* 球化处理
2. hardening ['hɑ:dəniŋ] *n.* 硬化；淬火
3. tempering ['tempəriŋ] *n.* 回火处理

4. annealing *n.* 退火

5. normalizing ['nɔ:məlaiziŋ] *n.* 正火

6. ductile ['dʌktail] *adj.* 可拉长的，可延展的

7. brittle ['britl] *adj.* 易碎的，易裂的

8. quenching ['kwentʃiŋ] *n.* 淬火

9. In understanding the related areas of metalworking and methods used today, the mechanical properties of metals are of the utmost importance.

此句中，主语为the mechanical properties of metals，而in understanding the related areas of metalworking and methods used today为状语。

10. The maximum degree of hardness obtainable in steel by direct hardening is determined largely by the carbon content.

obtainable in steel by direct hardening做定语修饰hardness，主语为the maximum degree of hardness，谓语为is determined。

Section ③ Metalworking Machine Tool 金属加工机床

Machine tools use the principles found in simple tools. Broadly speaking, the purpose of all machine tools is to produce metal parts by changing the shape, size, or finish of a piece of material. The shape of a part made with a machine tool is limited, largely by the types of motion the tool can apply. The need to control such motion has led to the design of many different types of machine tools.

Standard machine tools are usually grouped in five basic classes: lathes, drilling machines, planers, milling machines, and grinding machines. Each of these machines can change the shape, size, or finish of a workpiece. This is done by removing material from it. Each of these machines operates on the basic principle of applying motion to the tool or the workpiece while the two are in contact. However, each machine tool differs from the others in the type used or the method by which it is applied. A brief description of the different classes of machine tools will be confined to the basic operating characteristics of each class. The engine lathe is a machine tool that produces a cutting action by rotating the workpiece against the cutting edge of the tool. The line of cut follows a curve. The path of consecutive cuts is, in most cases, a straight line.

The shape produced is basically cylindrical.

The drilling machine is a machine tool that produces the necessary cutting action by the rotation of a multiple-edged cutting tool. Consecutive cuts are obtained by advancing the tool into the workpiece in a direction perpendicular to the tool's rotation. The combination of a rotating and straight-line motion of the tool produces a circular hole in the workpiece.

The shaper and planer can be discussed together. Both of these machines can machine flat surfaces with a single-point tool. Straight-line, or reciprocating, motion is applied to both the work-piece and tool in these machines. Reciprocating means alternately move back and forth. Very similar cutting tools are used on these machines on operations that are basically the same. The principal difference is that in the shaper the workpiece is stationary. The cutting action is produced by the cutting tool reciprocating back and forth along the workpiece. In the planer, the cutting action is produced by the reciprocation of the workpiece. The sequence of cuts is obtained by reciprocating the tool. The cutting tool feeds automatically across the work-piece a regulated amount for each cutting stroke as the table reciprocates.

A milling machine provides a cutting action to a rotating tool. The sequence of cuts is obtained by reciprocating the workpiece in a straight line and taking straight cuts. The surface produced by the milling cutter will normally be straight in at least one direction. The tool used on a milling machine is a multiple-edged tool. The surface produced by this tool conforms to the contour of the cut-ting edges.

Grinding is one of the most accurate of all the basic machining methods. A grinder differs from other machines in that it uses a tool made of emery, carborundum, or similar materials. The wheel, made up of many tiny cutting points, cuts with the entire surface area that comes in contact with the material being ground. Grinders cut with a grinding action, removing material in the form of tiny particles. A grinder generates the cutting action by rotating the grinding wheel. It may also rotate the workpiece, reciprocate the workpiece, or reciprocate the tool-head. All of these movements (which may be combined) produce a sequence of cuts.

The use of machine tools has contributed much to our modern civilization. Machine tools are used to make parts for automobiles, railroad equipment, airplanes, ships, and even space vehicles. Modern machine tools have been responsible for today's mass production techniques. These mass production methods enable people to afford many products that have become a part of their lives.

Notes and Expressions

1. lathe [leið] *n.* 车床
2. drilling machine 钻床
3. planer ['pleinə] *n.* 刨机，刨床
4. milling machine 铣床
5. grinding machine 磨床
6. boring ['bɔ:riŋ] *n.* 钻孔；镗削
7. carborundum [ˌkɑ:bə'rʌndəm] *n.* 碳化硅；[机] 金刚砂
8. abrasive wheel 砂轮
9. contour ['kɔntuə] *n.* 轮廓，外形
10. consecutive [kən'sekjutiv] *adj.* 连续的；连贯的
11. reciprocating [ri'siprəkeitiŋ] *adj.* 往复的；交互的；摆动的

【Self Evaluation】/【自我评价】

1. ______________relieves internal stresses and strain, there are ____________types of it used in industry.

(A) Annealing, 2　　(B) Annealing, 3

(C) Quenching, 2　　(D) Quenching, 3

2. The ability to support weight without bending or breaking is__________.

(A) strength　　(B) hardness

(C) toughness　　(D) corrosion

3. The properties of metal may be classified in three categories: ______________ , and______________.

(A) chemical properties

(B) physical properties

(C) mechanical properties

(D) technical properties

4. __________of metal can changes the structures of metals by the application of heat.

(A) machining　　(B) drilling

(C) heat treatment　　(D) grinding

5. _____ is one of the most accurate of all the basic machining methods.

(A) Milling (B) Drilling

(C) Cutting (D) Grinding

6. Match the following terms or phrases.

numerical control machine tools	热处理
plain carbon steel	回火
mechanical properties	自由锻
toughness	热轧
heat treatment	碳素钢
tempering	铣床
full annealing	再结晶温度
casting	机械特性
smith forging	反射炉
milling machine	冷加工
rolling	韧性
recrystallization temperature	完全退火
reverberatory furnace	铸造
cold-working	数控机床

7. Use the following terms or words to complete the following sentences, and change the form if necessary.

metal toughness water normalize hydroxide lathe strength

1) The nonmetals are defined as those whose __________ form acids.

2) Alloy is a mixture which is combined by two or more __________ elements.

3) __________ is the metal's ability to bend without breaking, and this property is combined with__________, or ability to support weight without bending or breaking.

4) To harden low carbon and medium plain-carbon steels, __________ is generally used as a quench.

5) __________ is a process used to relieve the internal stresses due to hot-working, cold-working, and machining.

6) The engine __________is a machine tool that produces a cutting action by rotating the work-piece against the cutting edge of the tool.

【Extensive Reading】/【拓展阅读】

Elevator Type
电梯类型

According to the use of elevators, it can be divided into two main types: passenger elevator and freight elevator. According to the hoist mechanism, elevator can be divided into four groups: traction elevators, hydraulic elevators, pneumatic vacuum elevators and climbing elevator.

■ Passenger Elevator 客梯

A passenger elevator is designed to move people between a building's floors. The Equitable Building completed in 1870 in New York City was the first office building to have passenger elevators.

Sometimes passenger elevators are used as a city transport along with funiculars. For example, there is a 3-station underground public elevator in Yalta, Ukraine, which takes passengers from the top of a hill above the Black Sea on which hotels are perched, to a tunnel located on the beach below.

■ Freight Elevators 货梯

A freight elevator, or goods lift, is an elevator designed to carry goods, rather than passengers. Freight elevators are generally required to display a written notice in the car that the use by passengers is prohibited, though certain freight elevators allow dual use through the use of an inconspicuous riser.

Freight elevators are typically larger and capable of carrying heavier loads than a passenger elevator, generally from 2300 to 4500 kg.

- **Traction Elevator 曳引电梯**

Geared traction machines are driven by AC or DC electric motors. Geared machines use worm gears to control mechanical movement of elevator cars by "rolling" steel hoist ropes over a drive sheave which is attached to a gearbox driven by a high speed motor. These machines are generally the best option for basement or overhead traction use for speeds up to 500 ft/min (2.5 m/s).

Gearless traction machines are low speed (low RPM), high torque electric motors powered either by AC or DC. In this case, the drive sheave is directly attached to the end of the motor.

In each case, cables are attached to a hitch plate on top of the cab or may be "underslung" below a cab, and then looped over the drive sheave to a counterweight attached to the opposite end of the cables which reduces the amount of power needed to move the cab. The counterweight is located in the hoist-way and rides a separate railway system; as the car goes up, the counterweight goes down, and vice versa. This action is powered by the traction machine which is directed by the controller, typically a relay logic or computerized device that directs starting, acceleration, deceleration and stopping of the elevator cab. The weight of the counterweight is typically equal to the weight of the elevator cab plus 40%–50% of the capacity of the elevator. The grooves in the drive sheave are specially designed to prevent the cables from slipping. "Traction" is provided to the ropes by the grip of the grooves in the sheave, thereby the name. As the ropes age and the traction grooves wear, some traction is lost and the ropes must be replaced and the sheave repaired or replaced. Sheave and rope wear may be significantly reduced by ensuring that all ropes have equal tension, thus sharing the load evenly. Rope tension equalisation may be achieved using a rope tension gauge, and is a simple way to extend the lifetime of the sheaves and ropes.

- **Hydraulic Elevator 液压电梯**

Conventional hydraulic elevators. They use an underground cylinder, are quite common for low level buildings with 2–5 floors (sometimes but seldom up to 6–8 floors), and have speeds of up to 200 feet/minute (1 meter/second).

Holeless hydraulic elevators were developed in the 1970s, and use a pair of above ground cylinders, which makes it practical for environmentally or cost sensitive buildings with 2, 3, or 4 floors.

Roped hydraulic elevators use both above ground cylinders and a rope system, which combines the reliability of in ground hydraulic with the versatility of holeless hydraulic, even though they can serve up to 8–10 floors.

Chapter 4

Sheetmetal Working
钣金加工

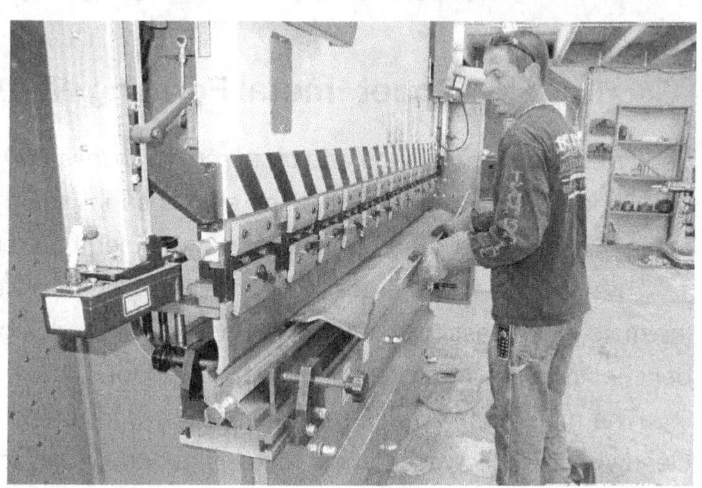

【Content Description】/【内容描述】

钣金在电梯中应用广泛，在许多装置上都存在钣金件，如电梯的轿厢、层门和轿门、检修箱和控制柜、导轨支架、护角板、限速器/张紧轮/反绳轮，扶梯的裙板和盖板等。钣金件的应用可以大幅度减少加工量，减轻零件的重量，提升整机的性能。

钣金的常用加工方法分成两大类：钣金成形和钣金切割。本章按照加工分类设置了两小节，分别展开叙述，讲解成形方法及设备。通过学习常见的钣金加工方法和工艺，了解和掌握钣金常见的分类、参数和设备的英语词汇，英语中成形方式定义的表达方法和一些加工方式的叙述规律和常用句型。

【Related Knowledge】/【知识准备】

钣金既可用来描述一种加工方式，也可以用来代指某类材料。钣金加工是一种针对均匀金属薄板进行加工的工艺方法，通过移除材料或材料变形得到预期形状的零件，包括冲裁、弯曲、剪切、拉伸以及焊接、铆接、表面处理等。关于钣金的厚度并没有明确定义，一般认为厚度0.15 ~ 6mm金属片材属于钣金范畴，比这更薄的材料称为箔，更厚的材料称为板。

钣金工厂最重要的加工方式是剪切、冲/切、折弯、焊接和表面处理。一般来说，钣金工厂基本设备包括剪板机、数控冲床、折弯机等。各种辅助设备包括开卷机、校平机、去毛刺机、点焊机等。常用的材料包括普通冷轧板、镀锌钢板和不锈钢。

【Section Implement】/【章节内容】

Section ❶ Sheet-metal Forming 钣金成型

Read the following passage and try to retell with less 200 words.

Sheet metal forming processes are those in which force is applied to a piece of sheet metal to modify its geometry rather than remove any material.

The applied force stresses the metal beyond its yield strength, causing the material to plastically deform, but not to fail. By doing so, the sheet can be bent or stretched into a variety of complex shapes.

The material thickness that classifies a work piece as sheet metal is not clearly defined. However, sheet metal is generally considered to be a piece of stock between 0.006 and 0.25 inches thick. A piece of metal much thinner is considered to be “foil” and any thicker is referred to as a “plate” .

■ Bending 折弯

Bending is a metal forming process in which a force is applied to a piece of sheet metal, causing it to bend at an angle and form the desired shape. A bending operation causes deformation along one axis, but a sequence of several different operations can be performed to create a complex part.

Bend parts can be quite small, such as a bracket, or up to 20 feet in length, such as a large enclosure or chassis. A bend can be characterized by several different parameters, shown in the figure4-1.

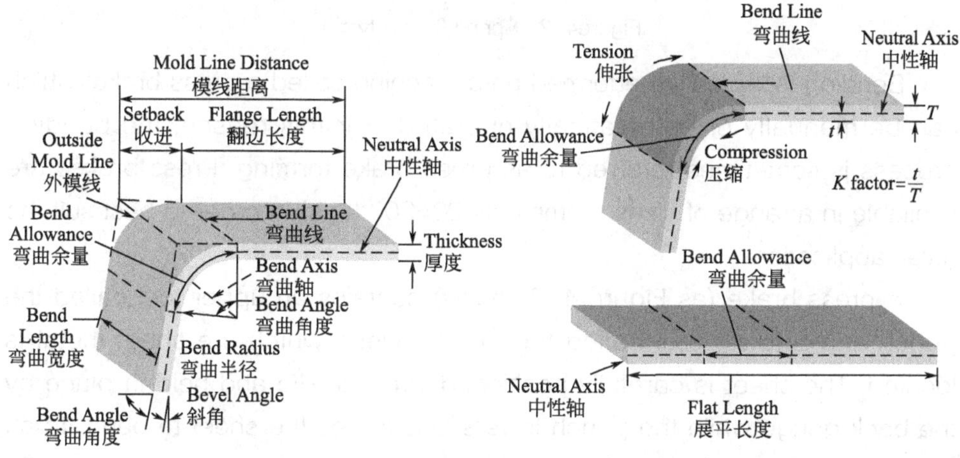

Figure4-1 Sheet-metal Bending（钣金折弯）

The act of bending results in both tension and compression in the sheet metal. The outside portion of the sheet metal will undergo tension and stretch to a great length, while the inside portion experiences compression and shortens. The neutral axis is the boundary line inside the sheet metal, along which no tension or compression forces are present. As a result, the length of this axis remains constant.

The changes in length to the outside and inside surfaces can be related to the original flat length by two parameters, the bend allowance and K factor, which are defined in figure4-2.

When bending a piece of sheet metal, the residual stress in the material will cause the sheet to spring-back slightly after bending operation. Due to this elastic recovery, it is necessary to over-bend the sheet a precise amount to achieve the desired bend radius and bend angle. The final bend radius will be greater than initially formed and the final bend angle will be smaller. The ratio of the final bend angle to the initial bend angle is defined as the spring-back factor, Ks. The amount of spring-back depends upon several factors, including the material, bending operation, and the initial bend angle and bend radius.

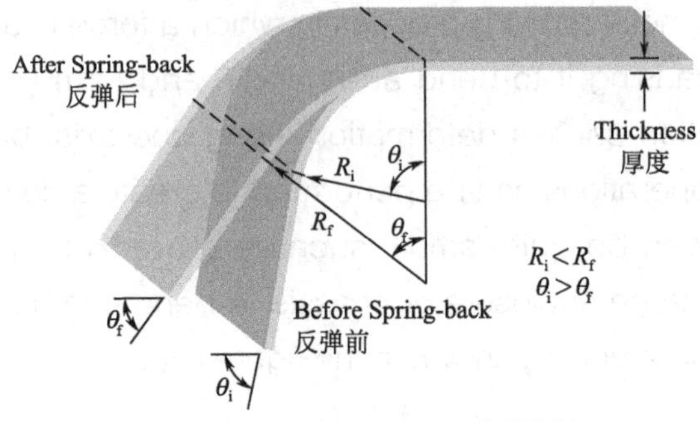

Figure4-2 Spring-back(反弹)

Bending is typically performed on a machine called a press brake, which can be manually or automatically operated. For this reason, the bending process is sometimes referred to as press brake forming. Press brakes are available in a range of sizes (commonly 20-200 tons) in order to best suit the given application.

A press brake (as Figure 4-3 shows) contains an upper tool called the punch and a lower tool called the die, between which the sheet metal is located. The sheet is carefully positioned over the die and held in place by the back gauge while the punch lowers and forces the sheet to bend. In an automatic machine, the punch is forced into the sheet under the power of a hydraulic ram. The bend angle achieved is determined by the depth to which the punch forces the sheet into the die. This depth is precisely controlled to achieve the desired bend.

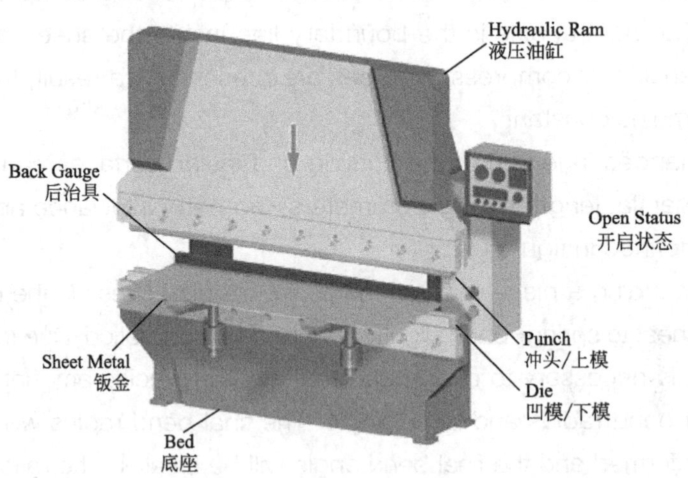

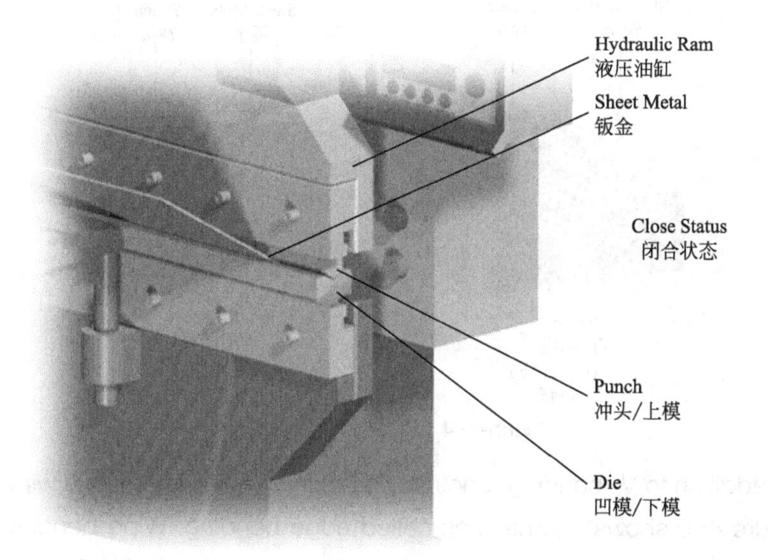

Figure4-3　Bending Machine/Press Brake（折弯机/弯板机）

Standard tooling is often used for the punch and die, allowing a low initial cost and suitability for low volume production. Custom tooling can be used for specialized bending operations but will add to the cost.

The tooling material is chosen based upon the production quantity, sheet metal material, and degree of bending. Naturally, a stronger tool is required to endure larger quantities, harder sheet metal, and severe bending operations. In order of increasing strength, some common tooling materials include low carbon steel, tool steel, and carbide steel.

While using a press brake and standard die sets, there are still a variety of techniques that can be used to bend the sheet. The most common method is known as V-bending, in which the punch and die both are "V" shaped. The punch pushes the sheet into the "V" shaped groove in the V-die, causing it to bend.

If the punch does not force the sheet to the bottom of the die cavity, leaving space or air underneath, it is called "air bending"(as Figure 4-4 shows). As a result, the V-groove must have a sharper angle than the angle being formed in the sheet. If the punch forces the sheet to the bottom of the die cavity, it is called "bottoming". This technique allows for more control over the angle because there is less spring back. This value is referred to as the die ratio and is equal to the die opening divided by the sheet thickness.

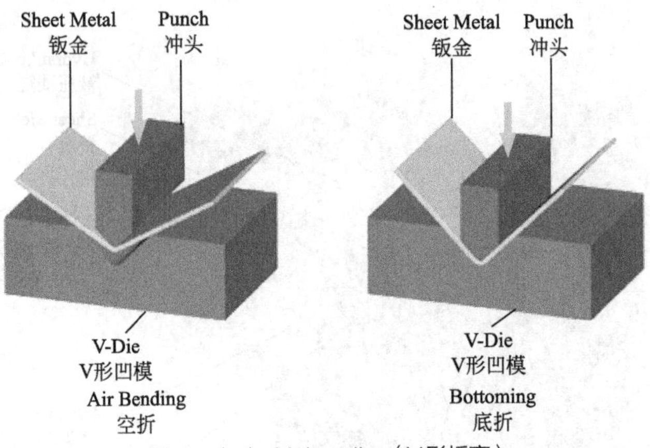

Figure4-4　V-bending（V形折弯）

In addition to V-bending, another common bending method is wipe bending (as Figure 4-5 shows), sometimes called edge bending. Wipe bending requires the sheet to be held against the wipe die by a pressure pad. The punch then presses against the edge of the sheet that extends beyond the die and pad. The sheet will bend against the radius of the edge of the wipe die.

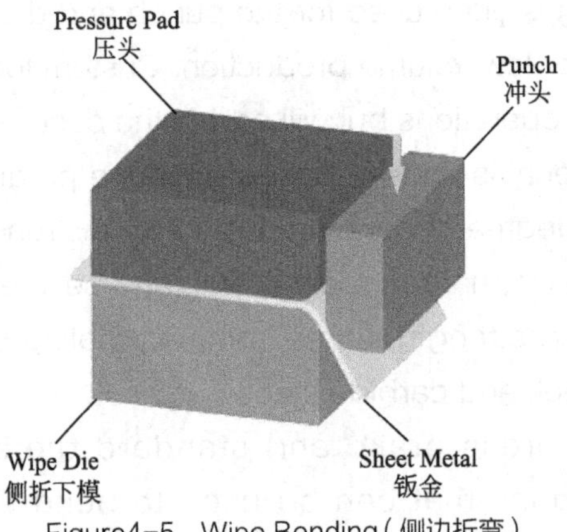

Figure4-5　Wipe Bending（侧边折弯）

Some design rules are listed as below:

1) Bend location — A bend should be located where enough material is present, and preferably with straight edges, for the sheet to be secured without slipping. Any features, such as holes or slots, located too close to a bend may be distorted. The distance of such features from the bend should be equal to at least 3 times the sheet thickness plus the bending radius.

2) Bend radius　— Use a single bend radius for all bends to eliminate additional tooling or setups. Inside bend radius should equal at least the sheet thickness, since too small inside bend radius may cause fracture during

bending;

3) Bend direction — Bending hard metals parallel to the rolling direction of the sheet may lead to fracture, so bending perpendicular to the rolling direction is recommended. In the case of manual bending, if the design allows, a slot can be cut along the bend line to reduce the manual force required.

■ Deep drawing 深拉成型

Deep drawn parts are characterized by a depth equal to more than half of the diameter of the part. These parts can have a variety of cross sections with straight, tapered, or even curved walls, but cylindrical or rectangular parts are most common.

Deep drawing (as Figure 4-6 shows) is most effective with ductile metals, such as aluminum, brass, copper, and mild steel. Examples of parts formed with deep drawing include automotive bodies and fuel tanks, cans, cups, kitchen sinks, and pots and pans.

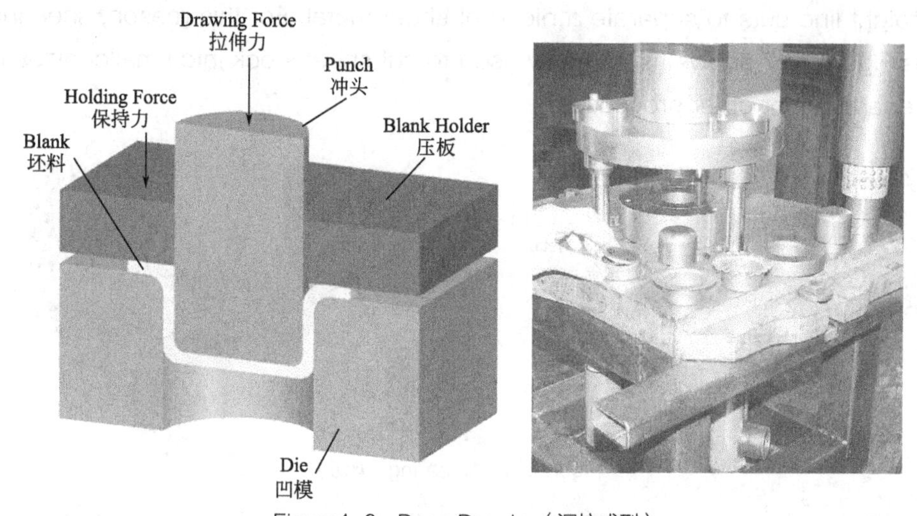

Figure4-6 Deep Drawing（深拉成型）

Notes and Expressions

1. ductile ['dʌktil] *adj.* 柔软的；易教导的；易延展的
2. cylindrical [si'lindrikəl] *adj.* 圆柱形的；圆柱体的
3. fracture ['fræktʃə] *n.* 破裂，断裂；骨折 *vi.* 破裂；折断
4. factor ['fæktə] *n.* 因素；要素；因子；系数
5. parameter [pə'ræmitə] *n.* 参数；参量
6. ram [ræm] *n.*（汽锤等的）撞锤；夯锤；（压力泵的）柱塞

Section ❷ Sheet-metal Cutting 钣金切割

Read the following passage and retell within 200 words.

Cutting processes are those in which a piece of sheet metal is separated by applying a great enough force to cause the material to fail. The most common cutting processes are performed by applying a shearing force, and are therefore sometimes referred to as shearing processes. When a great enough shearing force is applied, the shear stress in the material will exceed the ultimate shear strength and the material will fail and separate at the cut location.

▪ Shearing 剪切

A variety of cutting processes utilize shearing forces to separate or remove material from a piece of sheet stock in different ways. However, the term "shearing" by itself refers to a specific cutting process that produces straight line cuts to separate a piece of sheet metal. For this reason, shearing (as Figure 4-7 shows) is primarily used to cut sheet stock into smaller sizes in preparation for other processes.

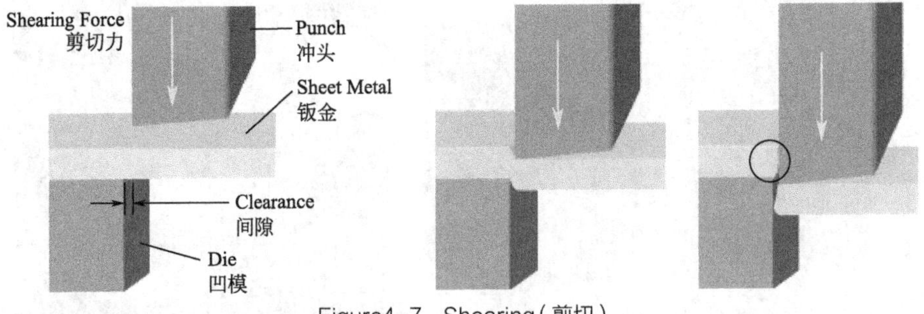

Figure4-7 Shearing(剪切)

This shearing force is applied by two tools, one above and one below the sheet. Whether these tools are a punch and die or upper and lower blades, the tool above the sheet delivers a quick downward blow to the sheet metal that rests over the lower tool. A small clearance is present between the edges of the upper and lower tools, which facilitates the fracture of the material. The size of this clearance is typically 2%-10% of the material thickness and depends upon several factors, such as the specific shearing process, material, and sheet thickness.

The shearing process is performed on a shear machine, often called a squaring shear or power shear, which can be operated manually (by hand or foot) and may be by hydraulic, pneumatic or electric power. A typical shear machine includes a table with support arms to hold the sheet, stops or guides to secure the sheet, upper and lower straight-edge blades, and a gauging device to precisely position the sheet. The sheet is placed between the upper and lower blade, which are then forced together against the sheet, cutting the material. In most devices, the lower blade remains stationary while the upper blade is forced downward.

▪ Blanking 下料

Blanking (as Figure 4-8 shows) is a cutting process in which a piece of sheet metal is removed from a larger piece of stock by applying a great enough shearing force. In this process, the piece removed, called the blank, is not scrap but rather the desired part. Blanked parts typically require secondary finishing to remove or smooth out burrs along the bottom edge.

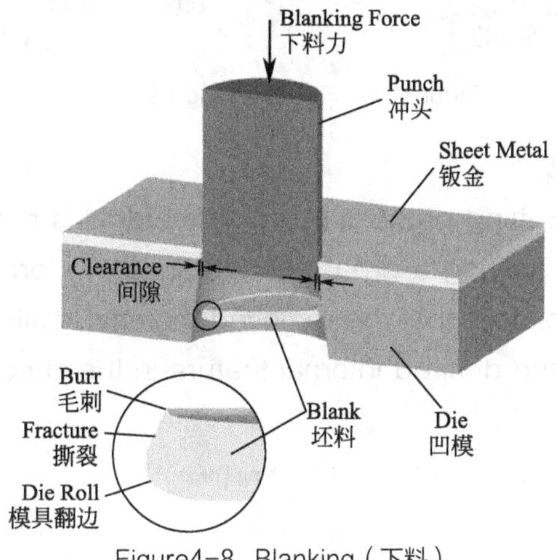

Figure4-8 Blanking（下料）

The blanking process requires a blanking press, sheet metal stock, blanking punch, and blanking die. The sheet metal stock is placed over the die. The die, instead of having a cavity, has a cutout in the shape of the desired part and must be custom made unless a standard shape is being formed. Above the sheet, resides the blanking punch which is a tool in the shape of the desired part.

Fine blanking (as Figure 4-9 shows) is a specialized type of blanking in which the blank is sheared from the sheet stock by applying 3 separate forces. This technique produces a part with better flatness, a smoother edge with minimal burrs, and tolerances as tight as ±0.001mm. As a result, high quality parts can be blanked that do not require any secondary operations. However, the additional equipment and tooling does add to the initial cost and makes fine blanking better suited to high volume production. Parts made with fine blanking include automotive parts, electronic components, cutlery, and power tools.

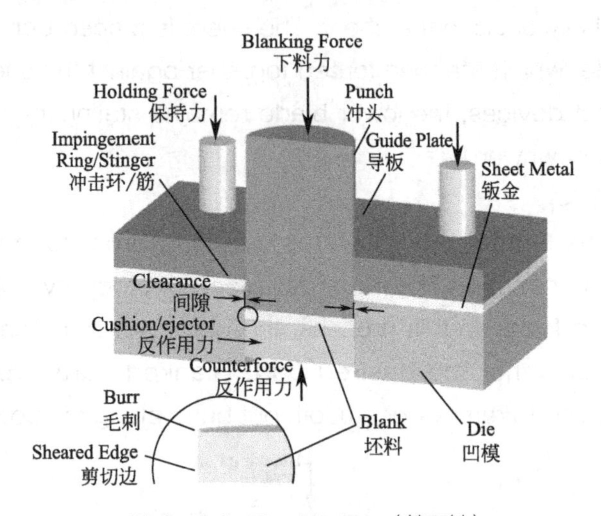

Figure4-9　Fine blanking（精下料）

▪ Punching 冲裁

Punching is a cutting process in which material is removed from a piece of sheet metal by applying a great enough shearing force. Punching is very similar to blanking except that the removed material, called the slug, is scrap and leaves behind the desired internal feature in the sheet, such as a hole or slot.

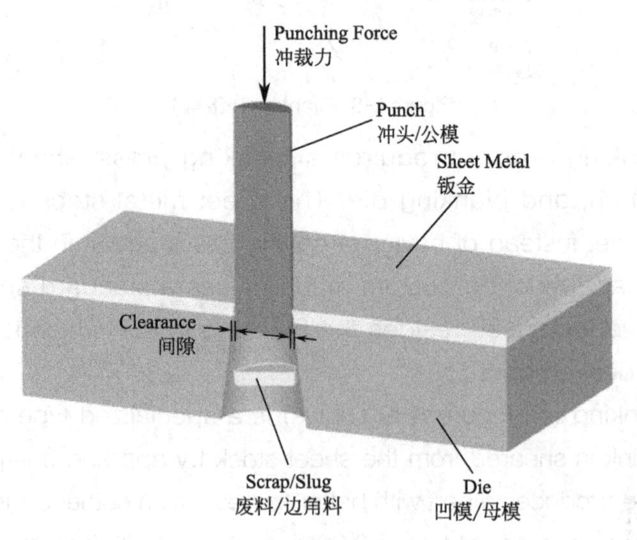

Figure4-10　Punching（冲裁）

Punching (as Figure 4-10 shows) can be used to produce holes and cutouts of various shapes and sizes. The most common punched holes are simple geometric shapes, like circle, square, rectangle, etc. The edges of these punched features will have some burrs from being sheared but are of fairly good quality. Secondary finishing operations are typically performed to attain smoother edges.

Notes and Expressions

1. scrap [skræp] *n.* 小块；碎片 碎屑；废料
2. finishing ['finiʃiŋ] *adj.* 最后的；终点的 *n.* 结束；结尾；精加工
3. flatness ['flætnis] *n.* 平面度；平整度；平坦
4. manually ['mænjuəli] *adv.* 手动地；用手
5. stress [stres] *n.* 压力；强调；紧张；应力；应激

【Self Evaluation】/【自我评价】

1. In blanking process, the piece removed is not scrap but rather the desired part. ( )

(A) True (B) False

2. Car wall is mainly composed of sheet metal and made by ________.

(A) cutting (B) deep drawing (C) bending (D) punching

3. Bending only results in tension in the sheetmetal. ( )

(A) True (B) False

4. Sheetmetal thickness is usually ________ mm.

(A) 2 ~ 10 (B) 0.006 ~ 0.25 (C) 0.15 ~ 6 (D) 0.15 ~ 0.6

5. Wipe bending and edge bending are ________.

(A) different (B) same (C) similar (D) opposite

6. Shearing is used to cut sheet stock into smaller sizes in preparation for other processes. ( )

(A) True (B) False

7. In most shear machine, the upper blade remains ________.

(A) rotary (B) stationary (C) move downward (D) move upward

8. Part after fine blanking requires secondary operation. ( )

(A) True (B) False

9. There is a clearance, ________ of the material thickness, between

punch and die.

(A) 2% ~ 5%　(B) 2% ~ 10%　(C) 5% ~ 10%　(D) 5% ~ 15%.

10. In fine blanking, the blank is sheared from the sheet stock by applying ______ separate forces.

(A) 1　(B) 2　(C) 3　(D) 4

【Extensive Reading】/【拓展阅读】

Basic Machining Techniques
基本机械加工方法

The importance of machining processes can be emphasized by the fact that every product we use in our daily life has undergone this process either directly or indirectly. In USA，more than S100 billions are spent annually on machining and related operations. A large majority (above 80%) of all the machine tools used in the manufacturing industry have undergone metal cutting.

The five basic techniques of machining metal include drilling and boring，turning，planing，milling and grinding. Variations of the five basic techniques are employed to meet special situations.

Drilling consists of cutting a round hole by means of a rotating drill. Boring，on the other hand，involves the finishing of a hole already drilled or cored by means of a rotating，offset，single-point tool. On some boring machines，the tool is stationary and the work revolves，on others，the reverse is true.

The lathe，as the turning machine is commonly called，is the father of all machine tools. The piece of metal to be machined is rotated and the cutting tool is advanced against it.

Planing metal with a machine tool is a process similar to planing wood with a hand plane. The essential difference lies in the fact that the cutting tool remains in a fixed position while the work is moved back and forth beneath it. Planers are usually large pieces of equipment; sometimes large enough to handle the machining of surfaces15to20feet wide and twice as long. A shaper differs from a planer in that the workpiece is held stationary and the cutting tool travels back and forth.

After lathes，milling machines are the most widely used for manufacturing applications. Milling consists of machining a piece of metal by bringing it into contact with a rotating cutting tool which has multiple cutting-edges. There are many types of milling machines designed for various kinds of work. Some of the shapes produced by milling machines are extremely simple, like the slots and flat surfaces produced by circular saws. Other shapes are more complex and may consist of a variety of combinations of flat and curved surfaces depending on the shape given to the cutting-edges of the tool and on the travel path of the tool.

Grinding consists of shaping a piece of work by bringing it into contact with a rotating abrasive wheel. The process is often used for the final finishing to close dimensions of a part that has been heat-treated to make it very hard. This is because grinding can correct distortions that may have resulted from heat treatment. In recent years, grinding has also found increased application in heavy-duty metal removal operations.

Polymer Process
聚合物加工

▪ Blow Molding 吹塑法

Blow molding is a manufacturing process that is used to create hollow plastic parts by inflating a heated plastic tube until it fills a mold and forms the desired shape.

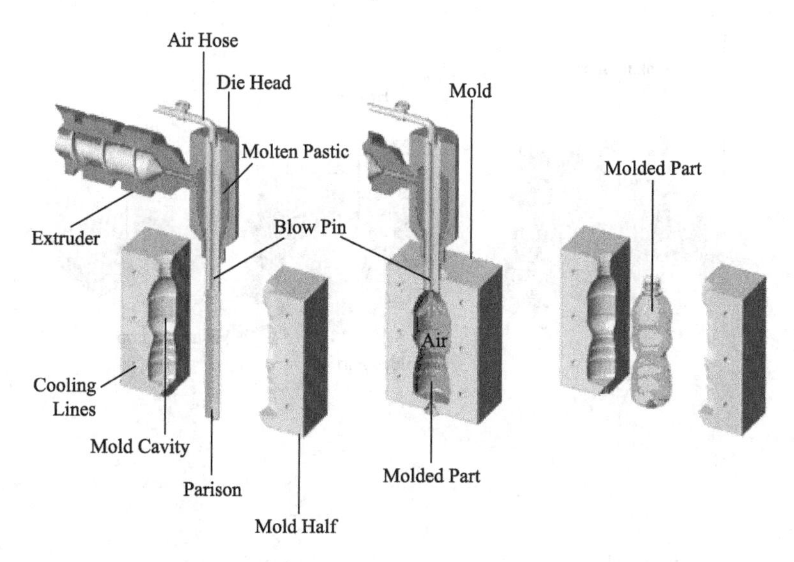

▪ Injection Molding 注射成型法

Injection molding is the most commonly used manufacturing process for the fabrication of plastic parts. A wide variety of products are manufactured using injection molding, which vary greatly in their size, complexity, and application.

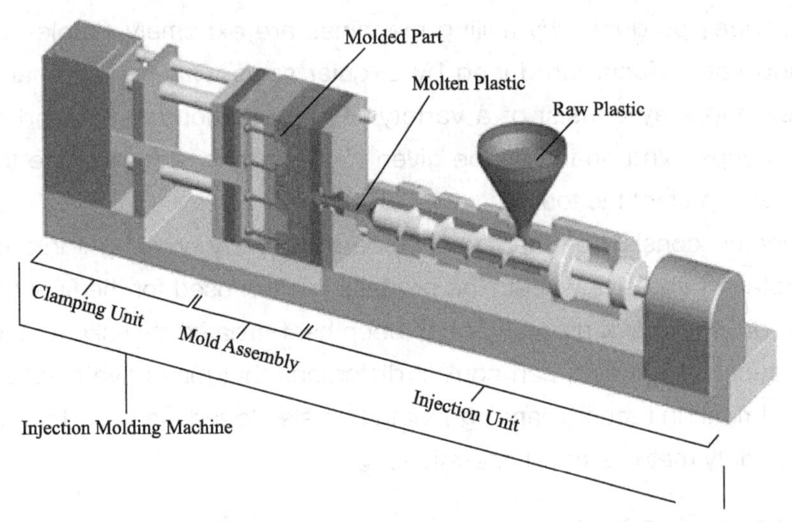

■ Thermoforming 热成型

Thermoforming describes the process of heating a thermoplastic sheet to its softening point, stretching it over or into a single-sided mold, and holding it in place while it cools and solidifies into the desired shape. The thermoplastic sheet is clamped into a holding device and heated by an oven using either convection or radiant heat until it is softened.

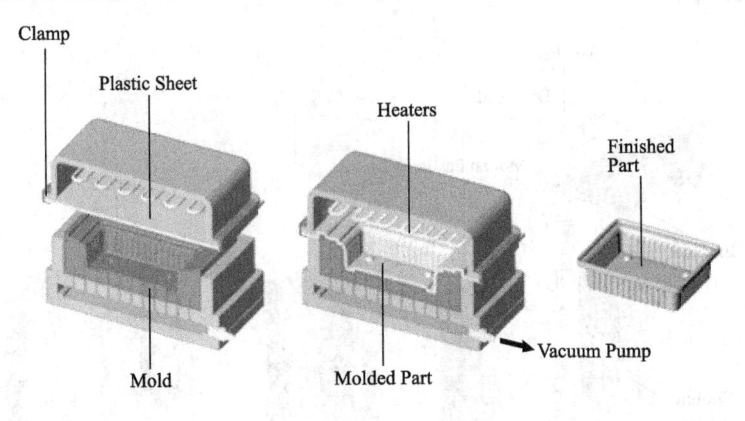

Chapter 5

Welding Technology
焊接技术

【Content Description】/【内容描述】

电梯的制造和安装过程中都会用到焊接，例如导轨支架和曳引机承重梁。上岗证和焊工证是从事电梯安装行业的重要资质和前提，掌握焊接相关的知识有利于安装的顺利进行和现场管理。电梯安装构件的焊接一般都采用手工电弧焊，具有机动灵活的特点，但又有质量不稳定的缺点，掌握不当，很容易产生焊缝缺陷和焊接变形。

本章设置了四个任务，分别介绍了常见的三种焊接方式（熔焊/压力焊/钎焊）、焊接后容易出现的缺陷及检测方法。通过本章学习应掌握常见焊接种类的英语词汇，原理和常用方法，提高对电梯结构的认知。

【Related Knowledge】/【知识准备】

焊接是通过加热、加压或两者并用，使工件的材质达到原子间的键合而形成永久性连接的工艺过程，焊接技术主要应用在金属母材上。金属焊接方法有40种左右，主要分为熔焊、压力焊和钎焊三大类。

熔焊是在焊接过程中将工件接口加热至熔化状态，不加压力完成焊接的方法。熔焊时，热源将待焊两工件接口处迅速加热熔化，形成熔池。熔池随热源向前移动，冷却后形成连续焊缝而将两工件连接成为一体。熔焊可以分为：电弧焊、电渣焊、气焊、电子束焊、激光焊等。最常见的电弧焊又可以进一步分为：手工电弧焊（焊条电弧焊）、气体保护焊、埋弧焊、等离子焊等。

压力焊是在加压条件下，使两工件在固态下实现原子间结合，又称固态焊接。常用的压力焊工艺是电阻对焊，当电流通过两工件的连接端时，该处因电阻很大而温度上升，当加热至塑性状态时，在轴向压力作用下连接成为一体。

钎焊是使用比工件熔点低的金属材料作钎料，将工件和钎料加热到高于钎料熔点、低于工件熔点的温度，利用液态钎料润湿工件，填充接口间隙并与工件实现原子间的相互扩散，从而实现焊接的方法。

具体工件焊接方法的选择必须经过设计者的仔细考虑，考虑因素包括被焊接金属的特性、连接方式、板材厚度、固定方式、可利用的设备、生产效率和焊接环境条件等。

【Section Implement】/【章节内容】

Section 1 Fusion Welding 熔焊

Read the following passages and describe what fuse welding is and list at least three types.

■ Arc Welding 电弧焊

Arc welding is a type of welding that uses a welding power supply to create an electric arc between an electrode and the base material to melt the metals at the welding point. They can use either direct or alternating current (DC or AC).

The most widely used of these methods is the shielded metal arc welding (SMAW) process，is also known as manual metal arc (MMA) welding (as Figure 5-1 shows) It is defined as an arc welding process wherein coalescence is produced by heating with an arc between a covered metal electrode and the workpiece. Shielding is obtained from decomposition of the electrode covering. Pressure is not used and filler metal is obtained from the electrode.

This welding technique is used in many fields, particularly in the manufacture of machinery, transportation equipment and piping systems.

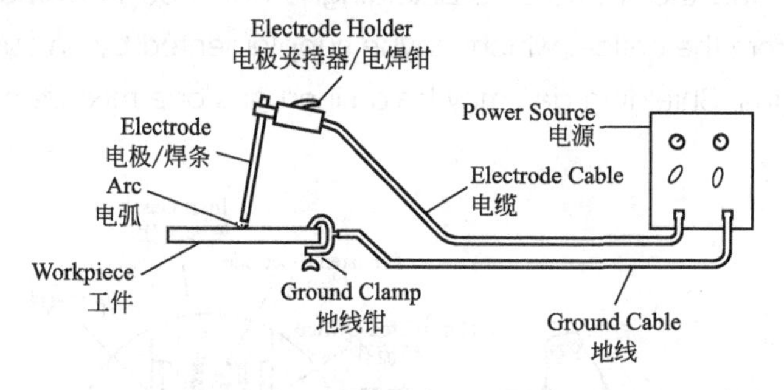

Figure5-1 Manual Metal Arc Welding（手工电弧焊）

Submerged arc welding (SAW) (as Figure 5-2 shows) is an arc welding process wherein coalescence is produced by heating with an arc between a bare metal electrode and the work. The arc is shielded by a blank of granular, fusible material on the workpiece. Pressure is not used and filler metal is obtained from the electrode and sometimes from a supplementary welding rod.

This method can be used in fully automated equipment where the feeds of both the electrode and granular flux care controlled. The method is also adaptable for semiautomatic equipment where the feed of the electrode and granular flux are controlled manually. Since the granular flux must cover the joint to be welded, this method is restricted to parts in horizontal position and is particularly suited for welding long straight joints. Also, fewer passes are needed to weld thick metal section than are usually required by shielded metal-arc welding.

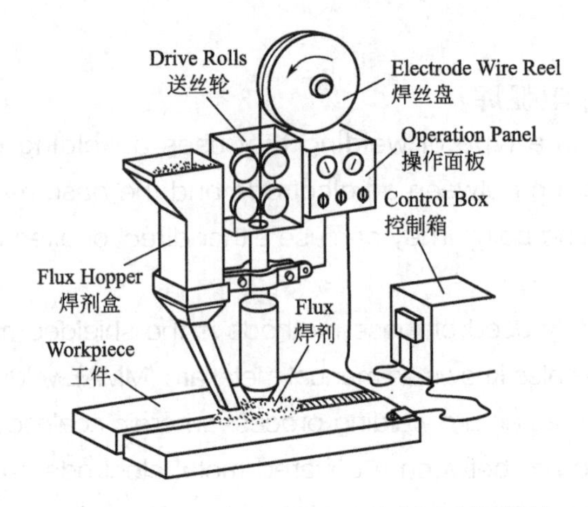

Figure5-2　Submerged Arc Welding（埋弧焊）

Plasma arc welding (PAW) (as Figure 5-3 shows) is arc welding process wherein coalescence is produced by heating with a constricted arc between an electrode and the workpiece. Shielding is obtained from the hot, ionized gas issuing from the orifice, which maybe supplemented by an auxiliary source of shielding gas. Shielding gas may be an inert gas or a mixture of gas.

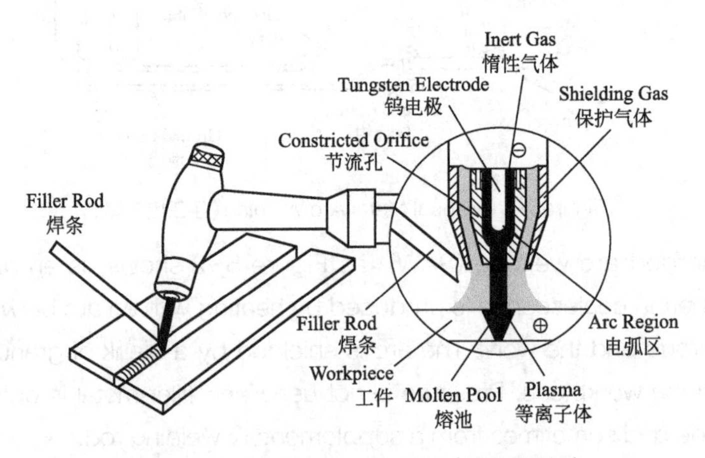

Figure5-3　Plasma Arc Welding（等离子弧焊）

▪ Gas Welding 气焊

Gas welding is a type of welding process that uses a flame that is fed by a pressurized gas fuel. Several different types of fuel gases can be used depending on application, availability, and cost. It is a very dangerous welding method since the gas used is flammable and combustible, what's more, in a pressure vessel. Gas welding has been actively used since the 19th century, and although it has been largely replaced by arc welding for many applications, it is still valued for its portability.

Oxyacetylene welding (OAW) is the one most frequently employed, it is commonly called oxy welding, or gas welding in the US. This method uses a mixture of oxygen and acetylene to produce heating. Fluxes may be used to reduce oxidation and to promote a better weld joint. This type of welding is suitable for both ferrous (including cast iron) and nonferrous metals and is capable of welding thick metal sections.

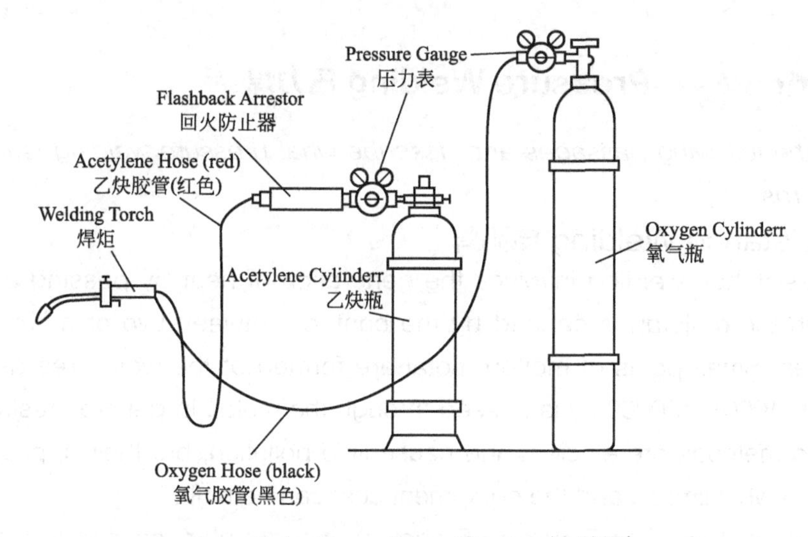

Figure5-4 Oxyacetylene Welding（氧炔焊）

The main advantage of the oxyacetylene welding process is that the equipment is simple, portable, and inexpensive. Therefore, it is convenient for maintenance and repair applications. However, due to its limited power density, the welding speed is very low and the total heat input per unit length of the weld is rather high, resulting in large heat-affected zones and severe distortion. The oxyacetylene welding process is not recommended for welding reactive metals such as titanium and zirconium.

Oxyhydrogen welding (OHW) (as Figure 5-4 shows) is used for low melting point metals such as aluminum, magnesium, and lead. Pressure gas welding (PGW) uses an oxyacetylene flame for a heat source but does not require a filler rod. Instead, fusion is obtained by applying pressure to the heated parts, either while being heated or after the parts are heated. This form of welding can be used for joining both ferrous and nonferrous metals.

Notes and Expressions

1. coalescence [ˌkəuə'lesns] *n.* 合并；联合；接合
2. granular ['grænjulə] *adj.* 颗粒的；粒状的
3. ionize ['aiənaiz] *vt.* 使电离，使离子化 *vi.* 电离，离子化
4. orifice ['ɔrifis] *n.* 孔，口，洞
5. inert [i'nə:t] *adj.* 惰性的；呆滞的；迟缓的；无效的
6. magnesium [mæg'ni:ziəm] *n.* 镁
7. oxidation [ˌɔksi'deiʃən] *n.*【化学】氧化；氧化作用

Section 2 Pressure Welding 压力焊

Read the following passages and describe what pressure welding is in less 100 words.

▪ Resistance Welding 电阻焊

Resistance welding involves the generation of heat by passing current through the resistance caused by the contact between two or more metal surfaces. Small pools of molten metal are formed at the weld area as high current (1000 - 100000 A) is passed through the metal. In general, resistance welding methods are efficient and cause little pollution, but their applications are somewhat limited and the equipment cost can be high.

Spot welding (as Figure 5-5 shows) is a popular resistance welding method used to join overlapping metal sheets of up to 3 mm thick. Two electrodes are simultaneously used to clamp the metal sheets together and to pass current through the sheets. The advantages of the method include efficient energy use, limited workpiece deformation, high production rates, easy automation, and no required filler materials.

Spot weld strength is significantly lower than with other welding methods, making the process suitable for only certain applications. It is used extensively in the automotive industry—ordinary cars can have several thousand spot welds made by industrial robots. A specialized process, called shot welding, can be used to spot weld stainless steel.

Like spot welding, seam welding (as Figure 5-6 shows) relies on two electrodes to apply pressure and current to join metal sheets. However, instead of pointed electrodes, wheel-shaped electrodes roll along and often feed the workpiece, making it possible to make long continuous welds.

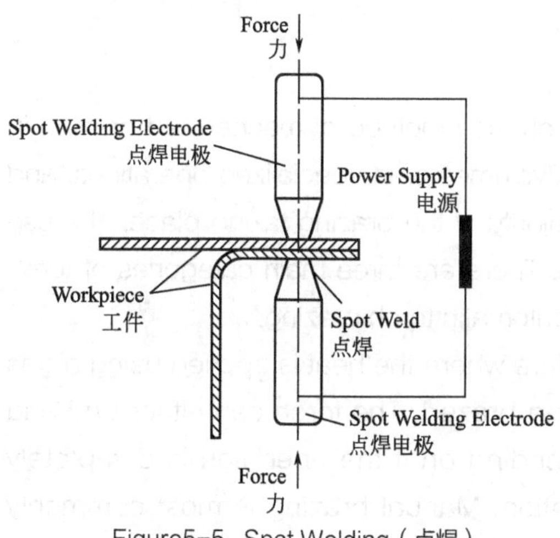

Figure5-5 Spot Welding（点焊）

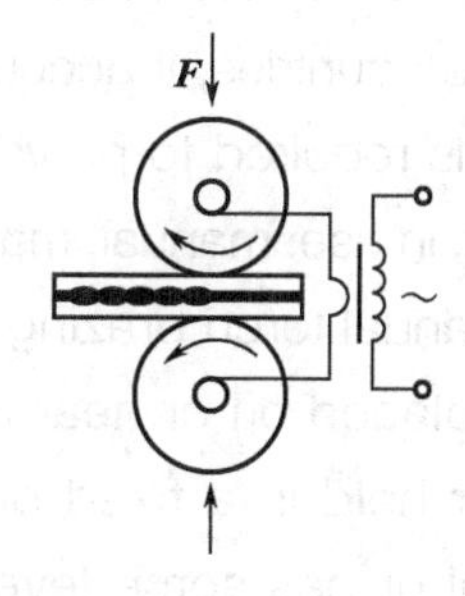

Figure5-6 Seam Welding（缝焊）

In the past, this process was used in the manufacture of beverage cans, but now its uses are more limited. Other resistance welding methods include butt welding, flash welding, projection welding, and upset welding.

Notes and Expressions

1. electrode [i'lektrəud] *n.* 电极；电焊条
2. clamp [klæmp] *vt.* 夹紧，固定住 *n.* 夹钳，螺丝钳

Section ❸ Braze Welding 钎焊

Read the following passages and describe what braze welding is and list at least three types.

Brazing has always been one of the most versatile and useful methods to join metals but until recent years, the equipment was bulky and expensive.

Brazing is a metal-joining process whereby a filler metal is heated above and distributed between two or more close-fitting parts by capillary action. The filler metal is brought slightly above its melting temperature while protected by a suitable atmosphere, usually a flux. It then flows over the base metal and is then cooled to join the workpieces together.

It is similar to soldering, except the temperatures used to melt the filler metal is above 450℃ , or, as traditionally defined in the United States, above 427 ℃ (800 ℉). Brazing creates an extremely strong joint, usually stronger than the base metal pieces themselves, without melting or deforming the components.

▪ Torch Brazing 火焰钎焊

Torch brazing is by far the most common method of mechanized brazing in use. It is best used in small production volumes or in specialized operations, and in some countries, it accounts for a majority of the brazing taking place. The use of flux is required to prevent oxidation. There are three main categories of torch brazing in use: manual, machine, and automatic torch brazing.

Manual torch brazing is a procedure where the heat is applied using a gas flame placed on or near the joint being brazed. The torch can either be hand held or held in a fixed position depending on if the operation is completely manual or has some level of automation. Manual brazing is most commonly used on small production volumes or in applications where the part size or configuration makes other brazing methods impossible. The main drawback is the high labor cost associated with the method as well as the operator skill required to obtain quality brazed joints.

Machine torch brazing is commonly used where a repetitive braze operation is being carried out. This method is a mix of both automated and manual operations with an operator often placing brazes material, flux and jigging parts while the machine mechanism carries out the actual braze. The advantage of this method is that it reduces the high labor and skill requirement of manual brazing. Automatic torch brazing is a method that almost eliminates the need for manual labor in the brazing operation.

▪ Furnace Brazing 炉内钎焊

Furnace brazing (as Figure 5-7 shows) is a semi-automatic process used widely in industrial brazing operations due to its adaptability to mass production and use of unskilled labor. There are many advantages of furnace brazing over other heating methods that make it ideal for mass production.

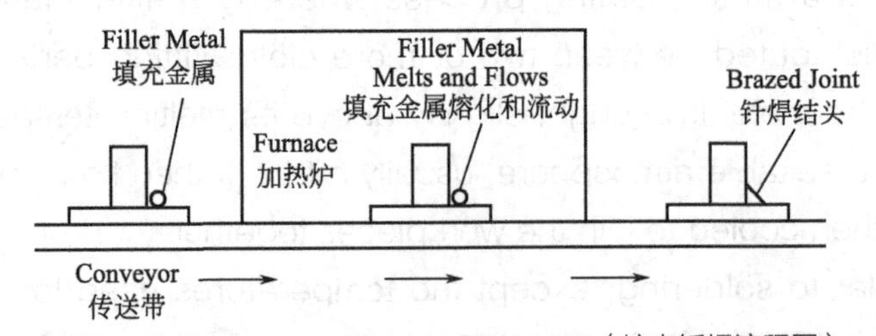

Figure5-7 Furnace Brazing Schematic（炉内钎焊流程图）

One main advantage is the ease with which it can produce large numbers of small parts that are easily jigged or self-locating. The process also offers the benefits of a controlled heat cycle and no need for post braze cleaning. Some other advantages include: low unit cost when used in mass production, close temperature control, and the ability to braze multiple joints at once. Furnaces are typically heated using either electric, gas or oil depending on the type of furnace and application.

However, some of the disadvantages of this method include: high capital equipment cost, more difficult design considerations and high power consumption.

Notes and Expressions

1. bulky ['bʌlki] *adj.* 体积大的；庞大的；笨重的
2. drawback ['drɔ:bæk] *n.* 缺点，不利条件；退税
3. mass production 大量生产；批量生产
4. joint [dʒɔint] *n.* 接头；接合点，接合处；接缝；关节；接头

Section 4 Welding Quality Control 焊接质量控制

Read the following passages and retell inspection method.

Ensuring that welders follow specific procedures is a crucial step in the overall welding quality system. Many distinct factors influence the strength of welds and the material around them, including the welding method, the amount and concentration of energy input, the weldability of the base material, filler material, and flux material, the design of the joint (as Figure 5-8 shows), and the interactions between all these factors.

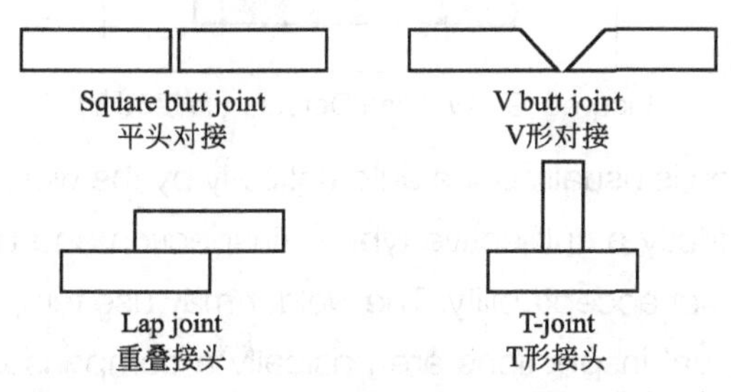

Figure5-8 Welding Joint Types（焊接搭接类型）

Types of welding defects include cracks, distortion, gas inclusions (porosity), non-metallic inclusions, lack of fusion, incomplete penetration, lamellar tearing, and undercutting (as shown in Figure 5-9). To test the quality of a weld, either destructive or nondestructive testing methods are commonly used to verify that welds are free of defects, have acceptable levels of residual stresses and distortion, and have acceptable heat-affected zone (HAZ) properties.

▪ Nondestructive Testing 无损检测

Nondestructive testing is a method of testing that does not destroy or impair the usefulness of a welded item. These tests disclose all of the common internal and surface defects that can occur when improper welding procedures are used. A large choice of testing devices is available and most of them are easier to use than the destructive methods, especially when working on large and expensive items.

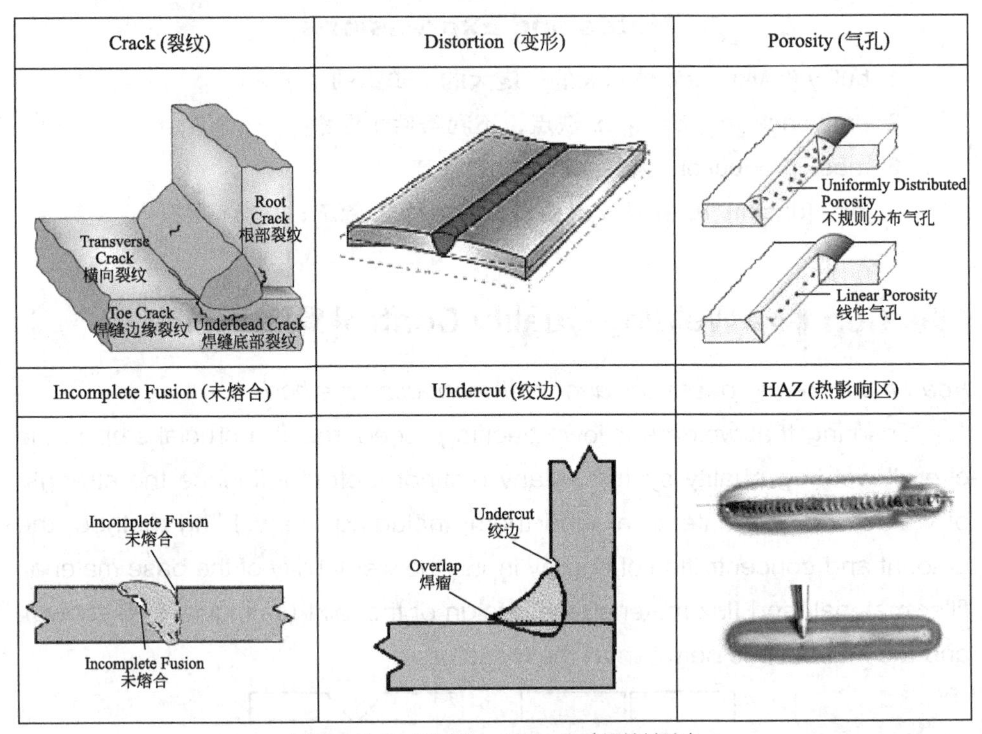

Figure5-9 Welding Defects（焊接缺陷）

Visual inspection is usually done automatically by the welder as he completes his welds. This is strictly a subjective type of inspection and usually there are no definite or rigid limits of acceptability. The welder may use templates for weld bead contour checks. Visual inspections are basically a comparison of finished welds with an accepted standard. This test is effective only when the visual qualities of a weld are the most important, otherwise, it is only a reference.

Magnetic particle inspection is most effective for the detection of surface or near surface flaws in welds. It is used in metals or alloys in which you can induce magnetism. While the test piece is magnetized, finely ground iron powder is applied. As long as the magnetic field is not disturbed, the iron particles will form a regular pattern on the surface of the test piece. When the magnetic field is interrupted by a crack or some other defect in the metal, the pattern of the suspended metal also is interrupted. The particles of metal cluster appear around the defect, making it easy to locate.

You can magnetize the test piece by either having an electric current pass through it, as shown in figure5−10(a), or by having an electric current pass through a coil of wire that surrounds the test piece, as shown in figure5−10(b). When an electric current flows in a straight line from one contact point to the other, magnetic lines of force are in a circular direction, as shown in figure 5−10(a). When the current flow is through a coil around the test piece, as shown in figure5−10(b), the magnetic lines of force are longitudinal through the test piece.

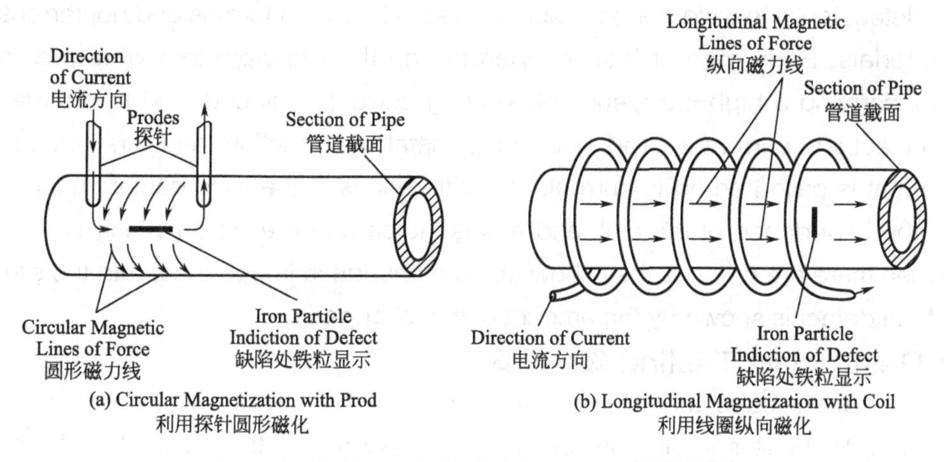

Figure5−10 Magnetic Particle Inspection（磁粉探伤）

Liquid penetrant methods are used to inspect metals for surface defects that are similar to those revealed by magnetic particle inspection. Unlike magnetic particle inspection, which can reveal subsurface defects, liquid penetrant inspection reveals only those defects that are open to the surface.

Radiographic inspection is a method of inspecting weldments by the use of rays that penetrate through the welds. X rays or gamma rays are the two types of waves used for this process. The rays pass through the weld and onto a sensitized film that is in direct contact with the back of the weld. When the film is developed, gas pockets, slag inclusions, cracks, or poor penetration will be visible on the film. Because of the danger of these rays, only qualified personnel are authorized to perform these tests.

Ultrasonic inspection of testing uses high-frequency vibrations or waves to locate and measure defects in welds. It can be used in both ferrous and nonferrous materials. This is an extremely sensitive system and can locate very fine surface and subsurface cracks as well as other types of defects, and all types of joints can be tested. This process uses high-frequency impulses to check the soundness of the weld. In a good weld, the signal travels through the weld to the other side and is then reflected back and shown on a calibrated screen. Irregularities, such as gas pockets or slag inclusions, cause the signal to reflect back sooner and will be displayed on the screen as a change in depth. When you use this system, most all types of materials can be checked for defects. Another advantage of this system is that only one side of the weld needs to be exposed for testing.

Eddy current is another type of testing that uses electromagnetic energy to detect faults in weld deposits and is effective for both ferrous and nonferrous materials. Eddy current testing operates on the principle that whenever a coil carrying a high-frequency alternating current is placed next to a metal, an electrical current is produced in the metal by induction, and this induced current is called an eddy current. The differences in the weld cause changes in the impedance of the coil, and this is indicated on electronic instruments. When there are defects, they show up as a change in impedance, and the size of the defect is shown by the amount of this change.

▪ Destructive Testing 破坏性测试

In destructive testing, sample portions of the welded structures are required. These samples are subjected to loads until they actually fail. The failed pieces are then studied and compared to known standards to determine the quality of the weld. The primary disadvantage of destructive testing is that an actual section of a weldment must be destroyed to evaluate the weld. This type of testing is usually used in the certification process of the welder.

The nick-break test is useful for determining the internal quality of the weld metal. This test reveals various internal defects, such as slag inclusions, gas pockets, lack of fusion, and oxidized or burned metal. To accomplish the nick-break test for checking a butt weld, you must first flame-cut the test specimens from a sample weld. Make a saw cut at each edge through the center of the weld. Next, place the saw-nicked specimen on two steel supports, as shown in figure5-11. Use a hammer to break the specimen by striking it in the zone where you made the saw cuts.

The free-bend test (as shown in Figure 5-12) is designed to measure the ductility of the weld deposit and the heat-affected area adjacent to the weld. Also it is used to determine the percentage of elongation of the weld metal. Ductility, you should recall, is that property of a metal that allows it to be drawn out or hammered thin.

The term tensile strength may be defined as the resistance to longitudinal stress or pull and is measured in pounds per square inch of cross section. Testing for tensile strength (as shown in Figure 5-13) involves placing a weld sample in a tensile testing machine and pulling on the test sample until it breaks.

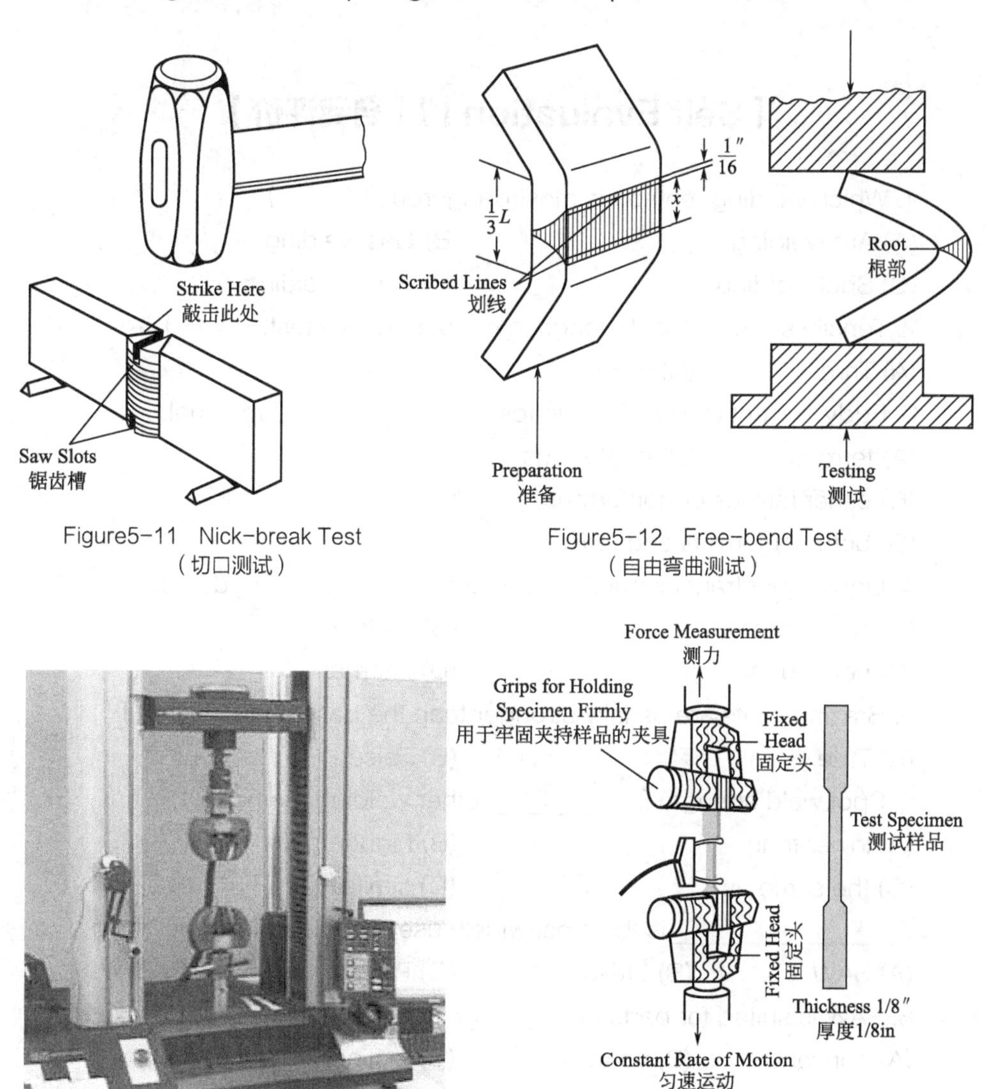

Figure5-11 Nick-break Test
（切口测试）

Figure5-12 Free-bend Test
（自由弯曲测试）

Figure5-13 Tensile Strength Test（抗拉强度试验）

Notes and Expressions

1. distinct [dis'tiŋkt] *adj.* 明显的；独特的；清楚的；有区别的
2. longitudinal [ˌlɔndʒi'tju:dinəl] *adj.* 长度的；纵向的（与transverse相对）；经度的
3. sensitize ['sensitaiz] *vt.* 使具感受力；使敏感；使感光
4. impedance [im'pi:dəns] *n.* 阻抗
5. tensile strength 抗张强度，抗拉强度，拉伸强度
6. weldment ['weldmənt] *n.* 焊件，焊接件

【Self Evaluation】/【自我评价】

1. Which welding method is most dangerous? (   )
(A) Arc welding (B) Gas welding
(C) Spot welding (D) Torch welding

2. Tensile strength test belongs to nondestructive test. (   )
(A) True (B) False

3. Eddy current test is effective for ________ material.
(A) ferrous (B) nonferrous
(C) either ferrous or nonferrous
(D) both nonferrous and ferrous

4. Liquid penetrant inspection can reveal ________ defects.
(A) subsurface (B) surface
(C) near surface (D) internal

5. Brazing joints are usually stronger than the base metal. (   )
(A) True (B) False

6. Spot weld strength is ________ other welding methods.
(A) lower than (B) higher than
(C) the same as (D) similar with

7. ________ is the most widely used method of arc welding.
(A) SAW (B) SMAW (C) PAW (D) OAW

8. SAW is suited for parts in ________ position.
(A) horizontal (B) vertical (C) inclined (D) any

9. Gas welding is still valued for its ________.
(A) portability (B) low cost (C) dangerous (D) high quality

10. Which one does belong to destructive testing method? (　　)

(A) magnetic particle inspection　　(B) ultrasonic inspection

(C) visual inspection　　(D) free-bend test

【Extensive Reading】/【拓展阅读】

Features of EST
科技英语特点

任何文章都可以归为特定类型的文体，根据文体不同，翻译方法也随之而异。科技文体崇尚严谨周密，概念准确，逻辑性强，行文简练，重点突出，句式严整，少有变化。而且常用前置性陈述，即在句中将主要信息尽量前置，通过主语传递主要信息。

科技文章的语言结构特色在翻译过程中如何处理，这是进行英汉科技翻译时需要探讨的问题。现分述如下。

一、大量使用名词化结构

《当代英语语法》在论述科技英语时提出，大量使用名词化结构是科技英语的特点之一。因为科技文体要求行文简洁、表达客观、内容确切、信息量大，强调存在的事实，而非某一行为。

【例】 The rotation of the earth on its own axis causes the change from day to night.

【译】 地球绕轴自转，引起昼夜的变化。

名词化结构 the rotation of the earth on its own axis 使复合句简化成简单句，而且使表达的概念更加确切严密。

【例】 Archimedes first discovered the principle of displacement of water by solid bodies.

【译】 阿基米德最先发现固体排水的原理。

名词化结构 of displacement of water by solid bodies，一方面简化了同位语从句，另一方强调 displacement 这一事实。

二、广泛使用被动语句

根据统计，科技英语中的谓语至少三分之一是被动态。这是因为科技文章侧重叙事推理，强调客观准确，因此尽量使用第三人称叙述，采用被动语态。

【例】 Attention must be paid to the working temperature of the machine.

【译】 应当注意机器的工作温度。

而很少说：

You must pay attention to the working temperature of the machine.

你们必须注意机器的工作温度。

此外，如前所述，科技文章将主要信息前置，放在主语部分。这也是广泛使用被动态的主要原因。试观察并比较下面短文的主语。

三、非限定动词

如前所述，科技文章要求行文简练，结构紧凑，为此，往往使用分词短语代替定语从句或状语从句；使用分词独立结构代替状语从句或并列分句；使用不定式短语代替各种从句；介词＋动名词短语代替定语从句或状语从句。这样既可缩短句子，又比较醒目。试比较下列各组句子。

【例】 A direct current is a current flowing always in the same direction.

【译】 直流电是一种总是沿同一方向流动的电流。

【例】 A body can more uniformly and in a straight line, there being no cause to change that motion.

【译】 如果没有改变物体运动的原因，那么物体将作匀速直线运动。

【例】 Vibrating objects produce sound waves ,each vibration producing one sound wave.

【译】 振动着的物体产生声波，每一次振动产生一个声波。

【例】 In communications, the problem of electronics is how to convey information from one place to another.

【译】 在通信系统中，电子学要解决的问题是如何把信息从一个地方传递到另一个地方。

【例】 Materials to be used for structural purposes are chosen so as to behave elastically in the environmental conditions.

【译】 结构材料的选择应使其在外界条件中保持其弹性。

【例】 There are different ways of changing energy from one form into another.

【译】 将能量从一种形式转变成另一种形式有各种不同的方法。

【例】 In making the radio waves correspond to each sound in turn, messages are carried from a broadcasting station to a receiving set.

【译】 使无线电波依次对每一个声音作出相应变化时，信息就由广播电台传递到接收机。

四、后置定语

大量使用后置定语也是科技文章的特点之一。常见的结构有以下五种。

▪ 1. 介词短语

【例】 The forces due to friction are called frictional forces.

【译】 由于摩擦而产生的力称为摩擦力。

【例】 A call for paper is now being issued.

【译】 征集论文的通知现正陆续发出。

▪ 2. 形容词及形容词短语

【例】 In this factory the only fuel available is coal.

【译】 该厂唯一可用的燃料是煤。

【例】 In radiation, thermal energy is transformed into radiant energy, similar in nature to light.

【译】 热能在辐射时，转换成性质与光相似的辐射能。

▪ 3. 副词

【例】 The air outside pressed the side in.

【译】 外面的空气将桶壁压得凹进去了。

【例】 The force upward equals the force downward so that the balloon stays at the level.

【译】 向上的力与向下的力相等，所以气球就保持在这一高度。

▪ 4. 单个分词，但仍保持较强的动词意义。

【例】 The results obtained must be cheeked.

【译】 获得的结果必须加以校核。

【例】 The heat produced is equal to the electrical energy wasted.

【译】 产生的热量等于浪费了的电能。

▪ 5. 定语从句

【例】 During construction, problems often arise which require design changes.

【译】 在施工过程中，常会出现需要改变设计的问题。

【例】 The molecules exert forces upon each other, which depend upon the distance between them.

【译】 分子相互间都存在着力的作用，力的大小取决于它们之间的距离。

【例】 Very wonderful changes in matter take place before our eyes every day to which we pay little attention.

【译】 我们几乎没有注意的很奇异的物质变化每天都在眼前发生。

【例】 To make an atomic bomb we have to use uranium 235, in which all the atoms are available for fission.

【译】 制造原子弹，我们必须用铀235，因为铀的所有原子都会裂变。

五、常用句型

科技文章中经常使用若干特定的句型，从而形成科技文体区别于其他文体的标志。例如It...that...结构句型，被动态结构句型，结构句型分词短语结构句型，省略句结构句型等。举例如下。

【例】 It is evident that a well lubricated bearing turns more easily than a dry one.

【译】 显然，润滑好的轴承比不润滑的轴承容易转动。

【例】 It seems that these two branches of science are mutually dependent and interacting.

【译】 看来这两个科学分支是相互依存，相互作用的。

【例】 It has been proved that induced voltage causes a current to flow in opposition to the force producing it.

【译】 已经证明，感应电压使电流的方向与产生电流的磁场力方向相反。

【例】 It was not until the 19th century that heat was considered as a form of energy.

【译】 直到十九世纪人们才认识到热是能量的一种形式。

【例】 Computers may be classified as analog and digital.

【译】 计算机可分为模拟计算机和数字计算机两种。

【例】 The switching time of the new-type transistor is shortened three times.

【译】 新型晶体管的开关时间缩短了三分之二（或缩短为三分之一）。

【例】 This steel alloy is believed to be the best available here.

【译】 人们认为这种合金钢是这里能提供的最好的合金钢。

【例】 Electromagnetic waves travel at the same speed as light.

【译】 电磁波传送的速度和光速相同。

【例】 Microcomputers are very small in size, as is shown in Fig.5.

【译】 如图5所示，微型计算机体积很小。

【例】 In water sound travels nearly five times as fast as in air.

【译】 声音在水中的传播速度几乎是在空气中传播速度的5倍。

【例】 Compared with hydrogen, oxygen is nearly 16 times as heavy.

【译】 与氢比较，氧的重量大约是它的16倍。

【例】 The resistance being very high, the current in the circuit was low.

【译】 由于电阻很大，电路中通过的电流就小。

【例】 Ice keeps the same temperature while melting.

【译】 冰在融化时，其温度保持不变。

【例】 An object, once in motion, will keep on moving because of its inertia.

【译】 物体一旦运动，就会因惯性而持续运动。

【例】 All substances, whether gaseous ,liquid or solid ,are made of atoms.

【译】 一切物质，不论是气态、液态，还是固态，都由原子组成。

六、长句

为了表述一个复杂概念，使之逻辑严密，结构紧凑，科技文章中往往出现许多长句。有的长句多达七八个词，以下即是一例。

【例】 The efforts that have been made to explain optical phenomena by means of the hypothesis of a medium having the same physical character as an elastic solid body led, in the first instance, to the understanding of a concrete example of a medium which can transmit transverse vibrations, and at a later stage to the definite conclusion that there is no luminiferous medium having the physical character assumed in the hypothesis.

【译】 为了解释光学现象，人们曾试图假定有一种具有与弹性固体相同的物理性质的介质。这种尝试的结果，最初曾使人们了解到一种能传输横向振动的具有上述假定的那种物理性质的发光介质。

Chapter 6

User Guide
用户指南

【Content Description】/【内容描述】

电梯企业大概分为整机和零部件两种类型。作为整机类型企业，经常需要采购一些大型的部件，如曳引机、限速器，需要阅读产品说明书，获取产品的参数、使用方法和注意事项。而作为零部件生产企业，需要撰写产品说明书，介绍产品的参数、使用方法和注意事项，让客户正确使用产品。

本章包含两节，曳引机的产品规格和使用指南。通过阅读这两个实际产品的资料，了解常见的参数和格式的常用英语词汇和句型。通过学习能阅读常见的曳引机、限速器等外购件的英文说明书等资料，了解说明书的内容和格式。

【Related Knowledge】/【知识准备】

曳引机：电梯曳引机是电梯的动力设备，又称电梯主机。功能是输送与传递动力使电梯运行。它由电动机、制动器、联轴器、(减速箱)、曳引轮、机架和导向轮及附属盘车手轮等组成。导向轮一般装在机架或机架下的承重梁上。附属盘车手轮有的固定在电动机轴上，也有平时挂在附近墙上，使用时再套在电动机轴上。

限速器：它随时监测控制着轿厢的速度，当出现超速情况时，即电梯额定速度的115%时，能及时发出信号，继而产生机械动作，切断供电电路，使曳引机制动。如果电梯仍然无法制动，则安装在轿厢底部的安全钳动作将轿厢强制制停。限速器是指令发出者，而安全钳是执行者，两者共同作用从而保障了安全电梯。

【Section Implement】/【章节内容】

Section ❶ Traction Machine Specification 曳引机规格书

Read the passage and find out the key parameters of BETM-06A.

In Line with Boda's philosophy to remain a step ahead in technology, Boda has developed a new model BETM-06A suitable for passengers' capacity up to 6 for cost effective solution to meet customer needs in this segment.

BETM-06A machine (as Figure 6-1 shows) is equipped with improved version of electric motor to provide smooth ride without jerks. All the critical components of the machine are manufactured in Boda's own manufacturing factory so as to get desired quality. Every machine is passed through rigorous inspection checks on Boda's latest measurement & testing equipment so as to conform that every machine meets designed norms.

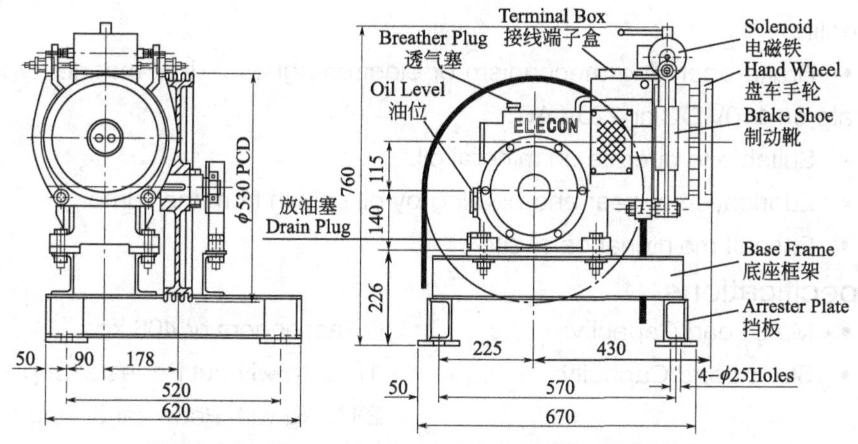

(a) With pedestal for travel height up to 33 mtr. (有底座，行驶高度33m)

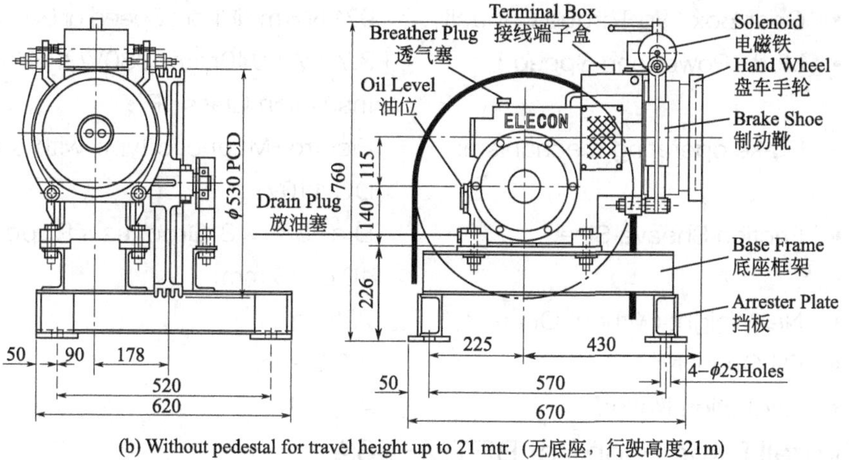

(b) Without pedestal for travel height up to 21 mtr. (无底座，行驶高度21m)

Figure6−1 BETM−06A Dimension (BETM−06A尺寸)

BETM−06A machine is available without & with pedestal support, as per customer's requirement.

▪ Salient Features

- Electric motor suitable for application on wider voltage range.
- Same unit can be used for DOL and VVVF application.
- Integral mounting of electric drive (stator and rotor) on high speed shaft.
- Compact dimensions and low weight.

- Monoblock housing of graded cast iron.
- Traction sheave of S. G. Iron.
- High speed shaft made out of hardened Alloy Steel.
- Worm wheel made out of special grade of Phosphor Bronze to achieve longer life.
- Brake operating mechanism of electromagnetic, dual magnet type to operate on 110V DC or 220V AC.
- Suitable to operate on mineral oil.
- Lubrication of gear and bearing by oil splash from the sump.
- Ease of maintenance at site.

▪ Specifications

- Max. Load Capacity: 6 Passengers or 408 kgs.
- Static Load Capacity: 1725 kg without Pedestal Support. 2375 kg with Pedestal Support.
- Rated Speed Maximum: 0.63 m/s.
- Gear Box O/p Torque Capacity: 921 N・m @ Input Speed of 940 rpm.
- Motor Power and Speed : 3.7 kW/ 940rpm, 380V/3PH/50Hz, Insulation Class ‘F’.
- Brake operating mechanism: Electro-Magnetic type with supply DC 110V.
- Friction Sheave Size: Dia. 530×3 Grooves×Rope Dia. 10 or 13 mm.
- Net Weight (without Oil) : 225 kg.
- Oil Quantity: 3.5 Litres.
- Reduction Ratio: 41 : 1.

▪ Overall Dimensions of BETM-06A

Notes and Expressions

1. rigorous ['rigərəs] *adj.* 严格的，严厉的；严密的；严酷的
2. pedestal ['pedistəl] *n.* 基架，基座；基础 *vt.* 支持；加座
3. salient ['seiljənt, -liənt] *adj.* 显著的；突出的；跳跃的 *n.* 凸角；突出部分
4. stator and rotor 定子和转子
5. monoblock *n.* 整体；直板；单块 *adj.* 整体的

Section ❷ Traction Machine Manual Book 曳引机使用手册

Read the passage and try to list the caution items.

Installation, Operation, Maintenance and Parts Manual for BETM-06A
BETM-06A安装、使用、维护及零件手册

▪ 1. Introduction

Congratulations for choosing BODA traction machine. This manual provides instructions for installation, operation and maintenance of the traction unit. Please go through this installation, operation and maintenance manual before you start using this product.

The satisfactory working of the unit depends on careful installation, correct grade of lubricant and good working conditions. Hence, it is most important to see that the installation of the unit is done according to the instructions given in this manual to ensure satisfactory working of the unit, and to ensure a long and trouble free service. Please note as under:

- The operations described in this manual must be carried out by qualified personnel equipped with proper tools and equipments.

• Before carrying out any maintenance work, the system must be put out of service.

• The serial number of the traction machine must always be quoted when ordering spare parts.

• All the traction machines are packed in wooden cases or crates.

• The traction machine must be unloaded from the lorry carefully with proper equipment.

• The condition of the machine should be checked at the time of taking delivery. If the machine is damaged, it should not be installed without clearance / approval from BODA.

• The Traction machine should be stored in its original packing in a dry place protected from the weather.

• If the Traction machine has already been unpacked, it must also be protected from dust.

• The client should protect the rotating shaft extension, sheave pulley with safety guards.

▪ 2. Effect of Long Storage

If the Traction Machine lies idle or is not put in to operation for a long period (more than 12 months from date of dispatch), the performance of the machine is likely to be affected and in that case Boda does not stand guaranty for satisfactory performance after installation. It is recommended to send back the traction machine to Boda's works to check the condition of bearings, oil seals, rubber parts and machine as a whole.

▪ 3. Installation & Operation

3.1 Handling

• Traction machine is supplied in completely assembled condition without oil. The sheave grooves are coated with anti corrosive agent, which is to be removed only by suitable solvent. In no case, it should be scraped or filed to remove the anti corrosive agent.

• The Traction machine should be lifted by making the use of the eyebolt or integrally cast lugs. These are meant for the lifting of the traction machine only and no accessories should be lifted along with traction machine.

• The Traction machine can be lifted with belts or chains, but care must be taken to avoid loading at certain critical parts.

3.2 Mounting

• Place the traction machine on its base and make sure that it is resting on all the contact points, including those of the external support if present, which correspond to the fixing holes. In case of uneven contact, keep some packings to ensure resting of base completely on floor.

• Tighten the bolts in diagonal order and fix the traction machine on the base.

• Ensure that the foundation has sufficient load carrying capacity.

3.3 Lubrication

• The oil used for lubrication is mineral base oil.

• Routine oil level check is very much necessary.

• Refilling instruction: With the Traction machine in standstill condition pour the oil through the hole until the centerline of the knob type oil level indicator is reached.

• Emptying instruction: With the Traction machine in standstill condition unscrew the drain plug at its base and drain the oil completely.

▪ Important Precautions

✧ First change of oil should be made after 5000 hours of operation.

✧ Subsequent oil change must be made after every 5000 hours of operation. The interval should not exceed 2 years.

✧ Cleanliness of oil is of prime importance and it is imperative to flush the gear unit with flushing oil before refilling. Fluid is to be drained of completely before filling the fresh oil.

✧ Oil of two different manufacturers should not be mixed in any case even though the same may be of equivalent grade.

3.4 Operation

• Using the hand wheel, make one complete turn manually of the drive pulley so that the oil distributes itself uniformly.

• Check the operation of the traction machine before attaching the ropes.

• The first 9–10 complete runs should be made at half load so as not to strain the gear.

• Once the traction machine has been checked, repeat the test with 1/4 load and successively with the cabin empty.

3.5 Brake Adjustment

All brakes (as Figure 6–2 shows) are factory set. It is recommended not to disturb factory setting unless it is must. However in case of brake requiring adjustment due to any reason following procedure should be followed:

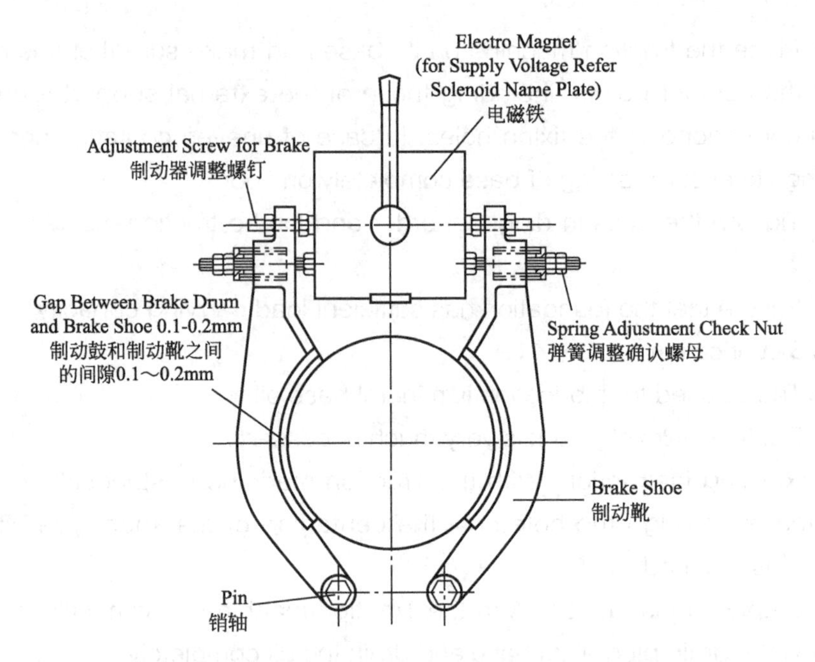

Figure6-2　Arrgt. for Electromagnet Brake（电磁制动器布局）

- The shoe should open with the least travel possible.
- Open the shoes with the brake lever.
- Screw or unscrew the adjustment screws until there is a gap of 0.1 to 0.2 mm between the shoes and the brake drum, check with the thickness gauge.
- The braking distance depends on the regulation of the springs, which must be adjusted from time to time.
- Check that the shoes open simultaneously during normal operation.
- Periodically check the brake shoes for wear.
- If the shoes are significantly worn, adjust the shoes in accordance with the instructions as above.
- The brake shoes must be replaced when the material is worn to 2mm thickness or less. Make sure that there are no oil traces on the brake drum or shoes.

■ 4. Maintenance

Carry out the following normal checks on the machine periodically. The frequency of the checks depends on the operation cycles of the machine. Every 6 months with operation cycles up to 2 hours a day, or every 3 months with operation cycles higher than 2 hours a day.

4.1 Axial Bearing Clearance Check

The clearance of the thrust bearing can be established visually by watching the axial movement of the brake drum with respect to the brake shoes during changes in direction. Contact our service department when such clearance becomes apparent on non-adjustable traction machines, so that the possibility of replacement can be ascertained.

4.2 Oil Change and Level Check

Refer 3.3 Lubrication.

4.3 Brake Shoe Wear Check

Refer 3.5 Brake Adjustment.

4.4 Oil Seal Check

Traction machine has static seals (without friction) and dynamic seals (with friction). Periodically check the traction machine for oil leaks and contact our service department if replacement of worn seal is necessary.

4.5 Friction Sheave Groove Wear Check

If the drive pulley grooves are worn out, the pulley must be replaced. Do not recondition the grooves without specific authorization.

4.6 Replacing Parts

Contact our service department along with machine serial number.

▪ 5. Instructions for storage, Installation and operation of Electric Motor

• When the motor is stored for a long time before starting always measure the insulation resistance using a 500 V Megger. If the value is very low (less than 5 mega ohms) on account of adverse weather condition (high humidity or moisture), the motor should be examined thoroughly.

• Machine must be installed as per the local electrical code/regulation by an authorized person. Motor must be protected against overload and short circuit conditions.

• Motor windings lead should be well tightened on motor terminal plate with bottom nut, washer and spring washer.

• Connect the motor in accordance with the connection (star or delta) given in the name plate and connection diagram given inside the terminal box cover.

• Power wire should be connected on motor terminal plate with ring type lugs. Conform that while connection of supply wire on motor, motor winding lead tightness is ok or not.

• While connecting the cables to motor terminals please ensure that the cable in properly secured and clamped and it does not exert any tension on terminals. A heavy load or tension by the cable can break the terminals.

- Connect earth terminal effectively for protection.
- The supply voltage should be same as given in the motor name plate or within the specified tolerance.
- Check motor current in no load condition and full load condition, it should be 10% less than written on name plate at rated voltage and should not exceed more than written on nameplate.
- Use motor thermister in control circuit to protect motor from overheating.

▪ Important

✧ The traction machines are designed and manufactured for usage in lifts and hoists and any other use other than this is to be considered as improper.

✧ The traction machines should not be used for the application other than given in the order specifications (Load, Speed, roping, mounting etc.).

✧ Required test and inspection must be carried out by qualified personnel.

✧ The maximum working temperature, measured on the housing, must not exceed 70℃ .

Notes and Expressions

1. crate [kreit] *n.* 板条箱；柳条箱；装货箱
2. lubricant ['lu:brikənt] *adj.* 润滑的 *n.* 润滑油；润滑剂
3. qualified ['kwɔlifaid] *adj.* 合格的；有资格的；能胜任的
4. lorry ['lɔ:ri;] *n.* 卡车；货车
5. dispatch [dis'pætʃ] *n.* 派遣；发货
6. eyebolt ['aibəult] *n.* 有眼螺栓；吊环螺栓
7. if present 如果存在
8. regulation [,regju'leiʃən] *n.* 规则；法规；管理；控制；调节；校准
9. earthing ['ə:θiŋ] *n.* [电] 接地（earth的现在分词）
10. thermister *n.* 热敏电阻

【Self Evaluation】/【自我评价】

1. Elevator traction machine BETM-06A can take ________ passengers one time.

(A) 4　　　(B) 6　　　(C) 8　　　(D) 10

2. BETM-06A brake operating mechanism is ________.

(A) Mechanical type　　(B) No　　(C) Electro-Magnetic type

3. The traction machines are designed and manufactured for usage in ________.

(A) lifts and hoists　　(B) escalators　　(C) moving walks

4. The diameter of rope used in BETM-06A friction sheave is ________mm.

(A) 8 or 10　　(B) 10 or 13　　(C) 6 or 8　　(D) 13 or 15

5. The rated speed max. of ETM045 is ________m/s.

(A) 0.5　　(B) 0.6　　(C) 0.55　　(D) 0.63

6. Traction machine and accessories can be lifted together.

(A) True　　(B) False　　(C) It depends　　(D) Maybe

7. The oil used for lubrication is mineral base oil.

(A) True　　(B) False　　(C) It depends　　(D) Maybe

8. The maximum working temperature measured on the housing must not exceed ________.

(A) 50℃　　(B) 70℃　　(C) 55℃　　(D) 65℃

9. BETM-06A should be maintained every ________ months with operation cycles higher than 2 hours day

(A) 3　　(B) 4　　(C) 5　　(D) 6

10. The Traction machine can be lifted with ________.

(A) belt only　　(B) belts or chains

(C) chains only　　(D) neither belt nor chain

【Extensive Reading】/【拓展阅读】

Character Introduction
人物介绍

■ Elisha Graves Otis 奥蒂斯

Elisha Graves Otis (August 3, 1811 - April 8, 1861) was an American industrialist, founder of the Otis Elevator Company, and inventor of a safety device that prevents elevators from falling if the hoisting cable fails. He worked on this device while living in Yonkers, New York in 1852, and had a finished product in 1854.

In 1853, Elisha Otis was a skilled 41-year-old mechanic in a bedstead factory. His employer asked the inventive mechanic to build a freight elevator to haul the bedsteads. In the process, Otis added a spring-and-ratchet safety to hold the platform if the hoist rope should break. It was reported to be the first elevator equipped with an automatic device to keep the elevator from falling.

▪ Andre-Marie Ampere 安培

André-Marie Ampere (20 January 1775-10 June 1836) was a French physicist and mathematician who is generally regarded as one of the main discoverers of electromagnetism. The SI unit of measurement of electric current, the ampere, is named after him.

Ampere was born in Lyon, France on 20 January 1775. He spent his childhood and adolescence at the family property at Poleymieux-au-Mont-d'Or near Lyon. His father began to teach him Latin, until he discovered the boy's preference and aptitude for mathematical studies.

Ampere's fame mainly rests on his establishing the relations between electricity and magnetism, and in developing the science of electromagnetism, or, as he called it, electrodynamics. During 1828, he was elected a foreign member of the Royal Swedish Academy of Science.

▪ Alessandro Antonio Volta 伏特

Alessandro Antonio Volta (1745-1827), an Italian physicist, invented the electric battery — which provided the first continuous

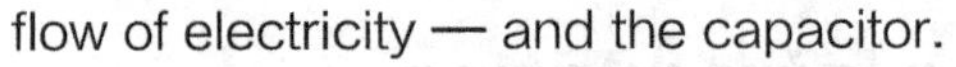

flow of electricity — and the capacitor.

Born in a noble family in Como, Italy, Volta was performing electrical experiments at age 18. his invention of the battery in 1796 revolutionized the use of electricity. The publication of his work in 1800 marked the beginning of electric circuit theory. Volta received many honors during his lifetime. The unit of voltage or potential difference, the volt, was named in his honor.

▪ Georg Simon Ohm 欧姆

Georg Simon Ohm (17 March 1789-6 July 1854) was a German physicist. As a high school teacher, Ohm began his research with the recently invented electrochemical cell, invented by Italian Count Alessandro Volta.

Using equipment of his own creation, Ohm determined that there is a direct proportionality between the potential difference (voltage) applied across a conductor and the resultant electric current. This relationship is now known as Ohm's law.

Chapter 7

Engineer Drawing
工程制图

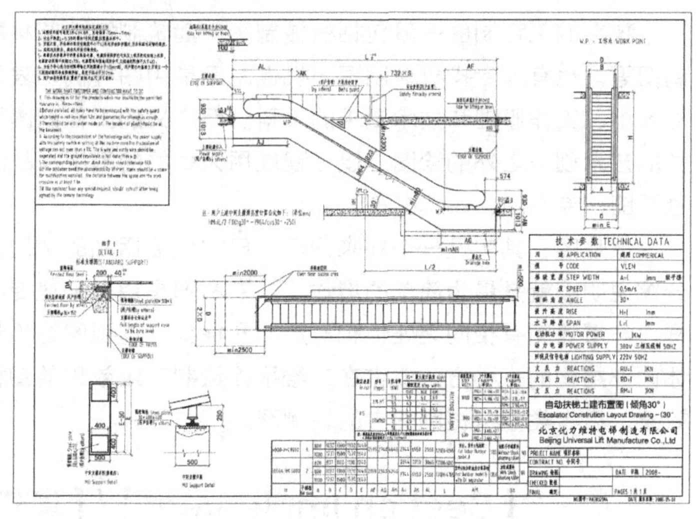

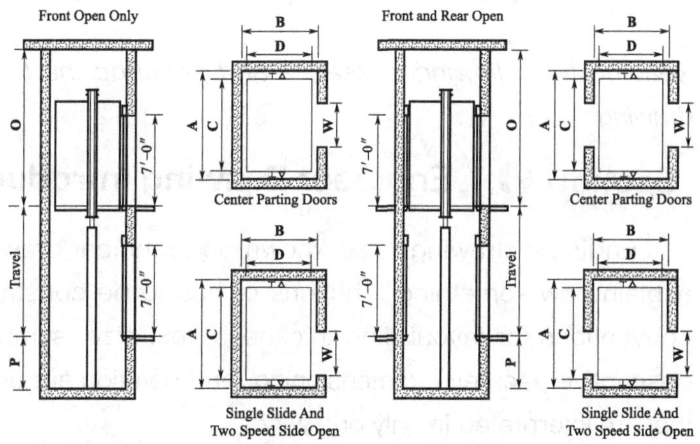

A=Hoistway Width
B=Hoistway Depth
C=Platform Width
D=Platform Depth
W=Door Width(Clear)
P=Pit Depth(Minimum 4′-0″)
O=Overheard:See Below for OH requirments

【Content Description】/【内容描述】

工程制图是工程技术中的一个重要组成部分。在高等工科课程中，它是一门重要的基础必修课，是研究工程图样的绘制和阅读的一门学科。通过学习，应掌握工程图的种类、机械制图的原理和方法。

本章共设置两节，工程制图简介和机械制图。第一节主要介绍工程制图的种类和特点，第二节主要介绍机械制图中的主要内容及其表达方法。通过本章学习，学生应能了解机械制图中图纸、投影、线条、字体等内容的表达方法。

【Related Knowledge】/【知识准备】

狭义的工程制图一般包括机械制图、建筑制图、园林制图、服装制图、家具制图等，也有人专指机械制图；机械制图是用图样确切表示机械的结构形状、尺寸大小、工作原理和技术要求的学科。而广义的工程制图是相对于绘画或素描之类的艺术创作之外的绘图，以工程应用为目的的图纸，故除上述类型外，还包含透视图、专利图、爆炸图等。

爆炸图，其实是一个外来词汇，日常生活用品的使用说明书中具有立体感的分解说明图就是最为简单的爆炸图。专利图是用来描述专利结构或主要特征的图样，与爆炸图有相同之处但相对严谨和抽象。透视图是拆除物体部分表面表达物体内部结构的一种方法或者在二维平面绘制三维效果的图样，前者类似爆炸图的效果，后者类似机械制图中的轴侧图。

【Section Implement】/【章节内容】

Reading the following passages and describe the feature of different type drawing.

Section 1 Engineer Drawing Introduction 工程制图简介

Engineer drawing, also known as technical drawing, which is used to explain how something functions or has to be constructed. Standards and conventions for layout, line thickness, text size, symbols, view projections, descriptive geometry, dimensioning, and notation are used to create drawings that are interpreted in only one way.

An Engineer drawing differs from a common drawing by how it is interpreted. A common drawing can hold many purpose and meanings, while an engineer drawing is intended to concisely and clearly communicate all needed specifications to transform an idea into physical form.

■ Architectural Drawing 建筑制图

Figure 7-1 Example of Architectural Drawing（建筑制图实例）

The art and design that goes into making buildings is known as architecture. To communicate all aspects of the design, detailed drawings are used. Architectural drawings (as Figure 7-1 shows) describe and document an architect's design.

An architectural drawing or architect's drawing is an engineer drawing of a building (or building project) that falls within the definition of architecture. Architectural drawings are used by architects and others for a number of purposes: to develop a design idea into a coherent proposal, to communicate ideas and concepts, to convince clients of the merits of a design, to enable a building contractor to construct it, as a record of the completed work, and to make a record of a building that already exists.

■ Mechanical Drawing 机械制图

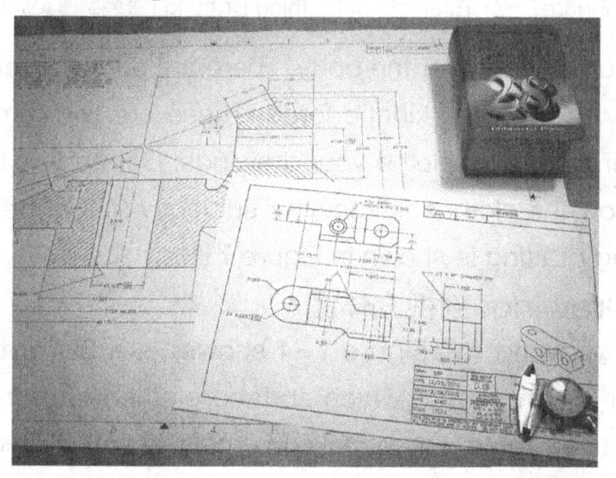

Figure 7-2 Example of Mechanical Drawing（机械制图实例）

A mechanical drawing (as Figure 7-2 shows), a type of engineer drawing, is used to fully and clearly define requirements for mechanical parts.

Mechanical drawing (the activity) produces engineering drawings (the documents). More than just the drawing of pictures, it is also a language—a graphical language that communicates ideas and information from one mind to another. Most especially, it communicates all needed information from the engineer who designed a part to the workers who will make it.

▪ Clothing Drafting 服装制图

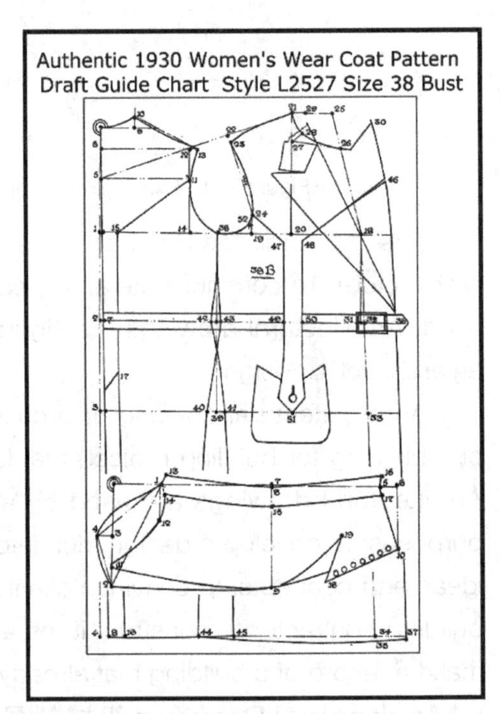

Figure 7-3 Example of Clothing Drafting (服装制图实例)

Clothes naturally flow over the body. There are a few factors that make this happen, namely gravity and kinetic force. These are two forces that you will want to consider when you clothe your drawings. Depending on the pose and these factors, how the clothes fall on your subject will vary between person to person. Chothing drafting is shown in Figure 7-3.

▪ Cutaway Drawing 剖视立体图

A cutaway drawing (as Figure 7-4 shows) is a 3D graphics, drawing, diagram and or illustration, in which surface elements a three-dimensional model are selectively removed, to make internal features visible.

Cutaway illustrations avoid ambiguities with respect to spatial ordering, provide a sharp contrast between foreground and background objects, and facilitate a good understanding of spatial ordering.

The location and shape to cut the outside object depends on many different factors, for example: the sizes and shapes of objects and personal taste, etc.

The cutaway view and the exploded view were minor graphic inventions of the Renaissance that also clarified pictorial representation. The term “Cutaway drawing” was already in use in the 19th century but became popular in the 1930s.

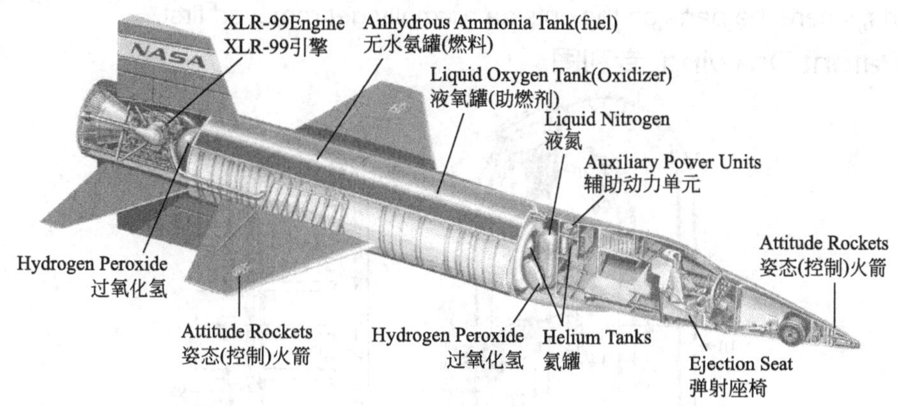

Figure 7-4　Example of Cutaway Drawing（剖视图实例）

▪ Exploded Drawing 爆炸图

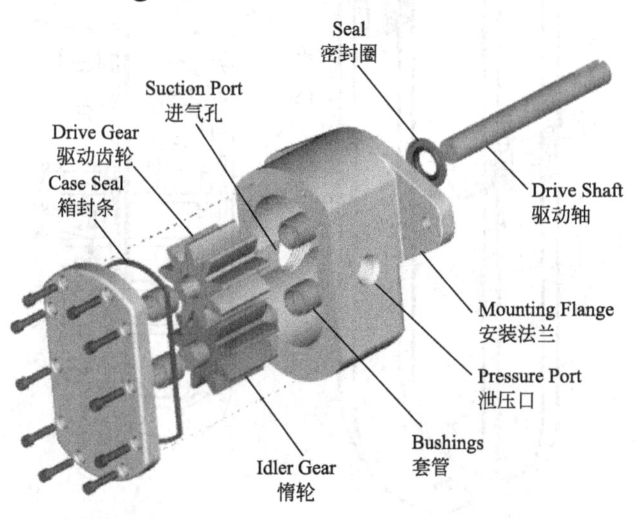

Figure 7-5　Example of Exploded Drawing（爆炸图实例）

An exploded view drawing (as Figure 7-5 shows) is engineer drawing of an object that shows the relationship or order of assembly of the various parts. It shows the components of an object slightly separated by distance, or suspended in surrounding space in the case of a three-dimensional exploded diagram. An object is represented as if there had been a small controlled explosion emanating from the middle of the object, causing the object's parts to be separated an equal distance away from their original locations.

An exploded view drawing can show the intended assembly of mechanical or other parts. In mechanical systems usually the component closest to the center is assembled first, or is the main part in which the other parts get assembled. This drawing can also help to represent disassembly of parts, where the parts on the outside normally get removed first.

▪ Patent Drawing 专利图

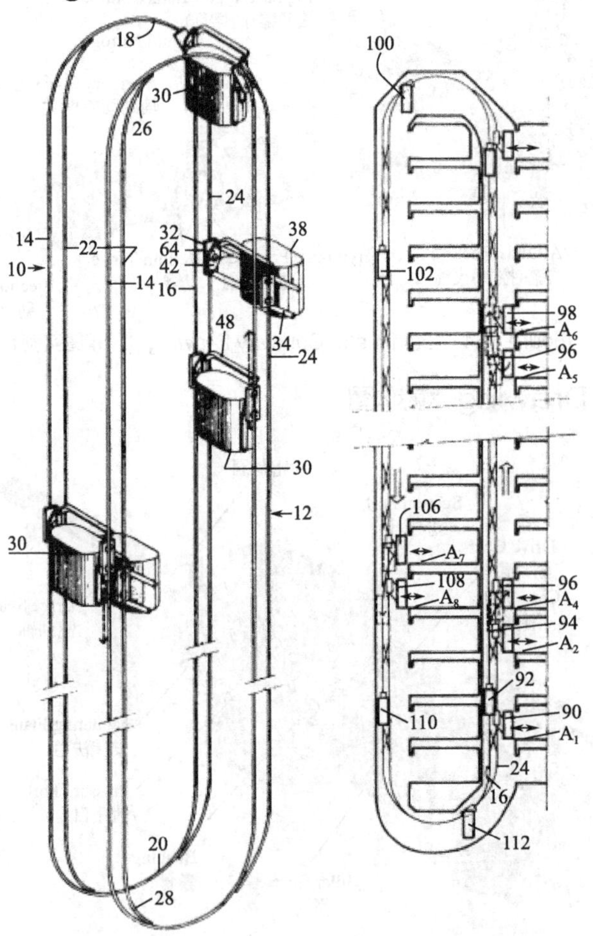

Figure 7-6 Example of Patent Drawing（专利图实例）

A patent application or patent may contain drawings, also called patent drawings (as Figure 7-6 shows), illustrating the invention, some of its embodiments (which are particular implementations or methods of carrying out the invention), or the prior art. The drawings may be required by the law to be in a particular form, and the requirements may vary depending on the jurisdiction.

The patent drawing can further contain a numbering of sheets of drawings, numbering of views, copyright notice, and security markings and corrections (durable and permanent).

Notes and Expressions

1. notation [nəu'teiʃən] *n.* 注释，批注；乐谱
2. descriptive geometry 画法几何(学)
3. renaissance [ri'neisəns] *n.* 新生；复活；文艺复兴
4. emanate ['eməneit] *vi.* 发出；流出；发源；散发；发射；放射；射出
5. jurisdiction [ˌdʒuəris'dikʃən] *n.* 司法；司法权；裁判权

Section 2 Mechanical Drawing 机械制图

Mechanical drawing (as Figure 7-7 shows) is a type of engineering drawing, created with the engineering discipline, and used to fully and clearly define requirements for engineer items. Engineering drawings are usually created in accordance with standardized conventions for layout, nomenclature, interpretation, appearance (such as typefaces and line styles), size, etc.

Its purpose is to accurately and unambiguously capture all the geometric features of a product or a component. The end goal of a mechanical drawing is to convey all the required information that will allow a manufacturer to produce that component.

"Make this part in accordance with the information indicated and to the dimension specified—any unauthorized deviations or errors are your responsibility." It is meant to convey the importance of the complete and proper dimensioning of engineering drawings. Careless dimensioning can lead to increase production costs or outright waste.

Almost all mechanical drawings communicate not only geometry (shape and location) but also dimensions and tolerances for those characteristics. Drawings convey the following critical information:

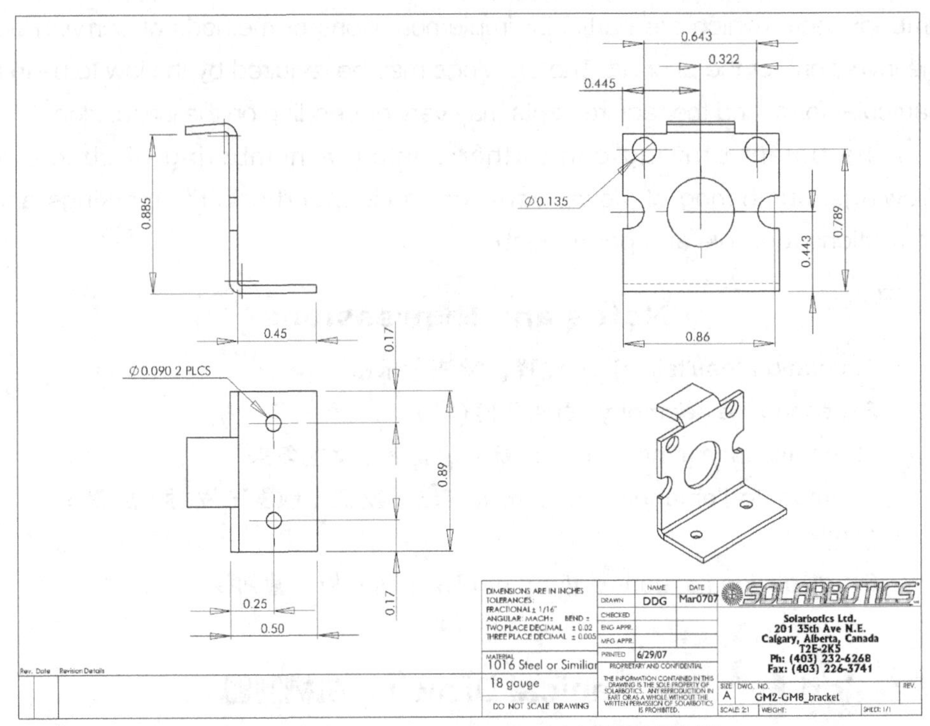

Figure 7-7 Bracket Part Drawing (支架零件图实例)

- Geometry – the shape of the object, represented as views, express how the object will look when it is viewed from various angles, such as front, top, side, etc.
- Dimensions – the size of the object is captured in accepted units.
- Tolerances – the allowable variations for each dimension.
- Material – represents what the item is made of.
- Finish – specifies the surface quality of the item, functional or cosmetic.

■ View Types and View Quantity 视图类型与数量

In most cases, a single view is not sufficient to show all necessary features, and several views are used. There are six basic views (as Figure 7-8 shows), but not all views are necessarily used. Generally only as many views are used as are necessary to convey all needed information clearly and economically.

The front, top, and right-side views are commonly considered the core group of views included by default, the determination of what surface constitutes the front, back, top, and bottom varies depending on the projection method used. Any combination of views may be used depending on the needs of the particular design.

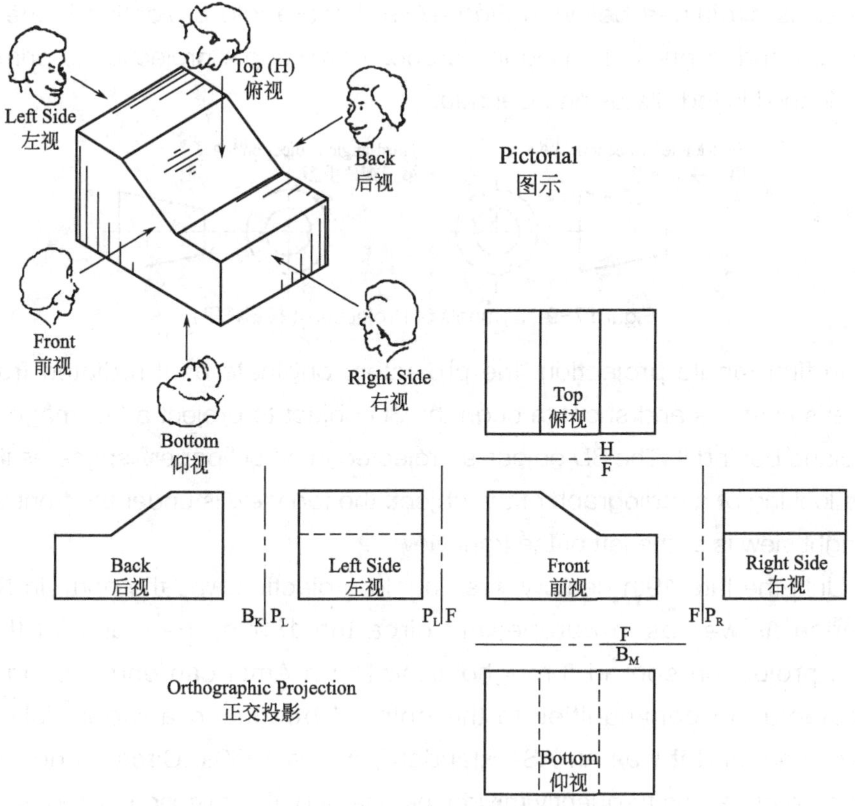

Figure 7-8 Principle Views（基本视图）

In addition to the 6 principal views (front, back, top, bottom, right side, and left side), any auxiliary views or sections may be included as serve the purposes of part definition and its communication. View lines or section lines (lines with arrows marked "A-A" , "B-B" , etc) define the direction and location of viewing or sectioning. Sometimes a note tells the reader in which zone(s) of the drawing to find the view or section.

Section views are projected views (either auxiliary or orthographic) which show a cross section of the source object along the specified cut plane. These views are commonly used to show internal features with more clarity than may be available using regular projections or hidden lines. In assembly drawings, hardware components (e.g. nuts, screws, washers) are typically not sectioned.

▪ Orthographic Projection 正交投影

The orthographic projection shows the object as it looks from the front, right, left, top, bottom, or back, and are typically positioned relative to each other according to the rules of either first-angle or third-angle projection. The origin and vector direction of the projectors (also called projection lines) differs, as explained below in Figure7-9. First-angle projection is the ISO standard and is primarily used in Europe. Third-angle projection is primarily used in the United States and Canada.

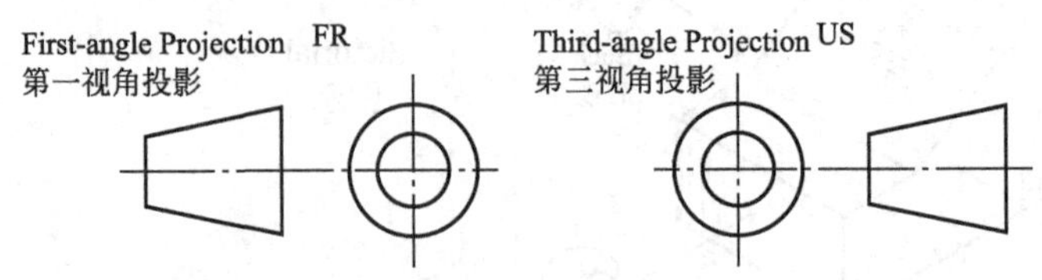

Figure 7-9 Symbols of Projection（投影符号）

In first-angle projection, the projectors originate as if radiated from a viewer's eyeballs and shoot through the 3D object to project a 2D image onto the plane behind it. The 3D object is projected into 2D "paper" space as if you were looking at a radiograph of the object: the top view is under the front view; the right view is at the left of the front view.

Until the late 19th century, first-angle projection was the norm in North America as well as in Europe; but circa the 1890s, the meme of third-angle projection spread throughout the North American engineering and manufacturing communities to the point of becoming a widely followed convention, and it was an ASA standard by the 1950s. Circa World War I, British practice was frequently mixing the use of both projection methods.

■ Basic Drafting Paper Sizes 基本图纸尺寸

The metric drawing sizes correspond to international paper sizes (as Figure 7-10 shows). These developed further refinements in the second half of the twentieth century, when photocopying became cheap. Engineering drawings could be readily doubled (or halved) in size and put on the next larger (or, respectively, smaller) size of paper with no waste of space.

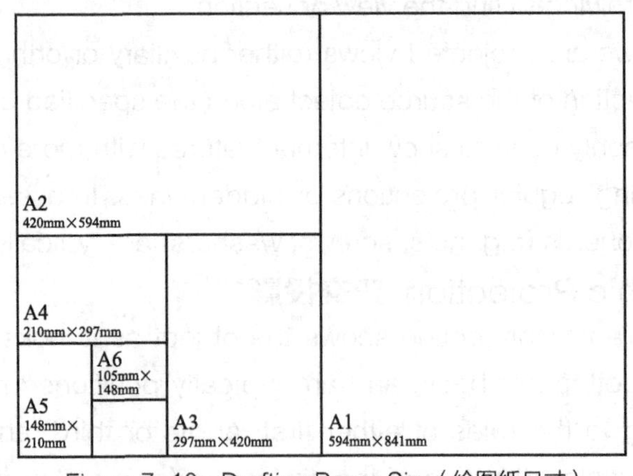

Figure 7-10 Drafting Paper Size（绘图纸尺寸）

All ISO paper sizes have the same aspect ratio, one to the square root of 2, meaning that a document designed for any given size can be enlarged or reduced to any other size and will fit perfectly. Given this ease of changing sizes, it is of course common to copy or print a given document on different sizes of paper, especially within a series, e.g. a drawing on A3 may be enlarged to A2 or reduced to A4.

▪ Drawing Methods 绘制方式

A sketch [as Figure 7-11(a) shows] is a quickly executed freehand drawing that is not intended as a finished work. In general, a sketch is a quick way to record an idea for later use. Architect's sketches primarily serve as a way to try out different ideas and establish a composition, especially when the finished work is expensive and time consuming.

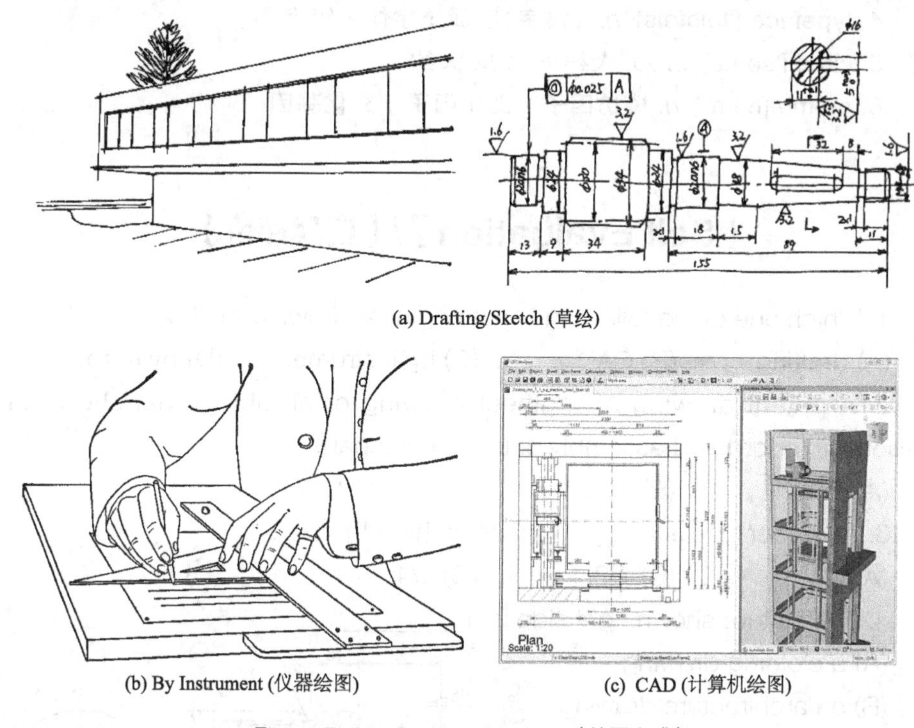

(a) Drafting/Sketch (草绘)

(b) By Instrument (仪器绘图)

(c) CAD (计算机绘图)

Figure 7-11 Drawing Methods（绘图方式）

Drawing by instrument [as Figure 7-11(b) shows] is to place a piece of paper (or other material) on a smooth surface with right-angle corners and straight sides—typically a drawing board. A sliding straightedge known as a T-square is then placed on one of the sides, allowing it to be slid across the side of the table, and over the surface of the paper. Compasses are used for drawing simple arcs and circles and triangular rule used for drawing straight lines.

Today, the mechanics of the drafting task have largely been automated and accelerated through the use of computer-aided design systems (CAD) [as Figure 7-11(c) shows]. There are two types of CAD systems used for the producing of engineer drawings, that is two dimensions (2D) drawing and three dimensions (3D) drawing. AutoCAD is the earliest and now the most popular software in 2D design, drafting, and modeling. The most well-known 3D software include Solidworks, ProE, UG and Catia.

Notes and Expressions

1. nomenclature [nəu'menklətʃə, 'nəumənkleitʃə] *n.* 命名法；术语
2. compass ['kʌmpəs] *n.* 指南针，罗盘；圆规 *vt.* 包围
3. triangular rule 三角定规；三角尺
4. typeface ['taipfeis] *n.* （铅字）字面；字体；铅字
5. circa ['sə:kə] *prep.* 大约于 *adv.* 大约
6. meme [mi:m] *n.* 模仿因子；文化因子；文化基因

【Self Evaluation】/【自我评价】

1. Which one of the following is not engineer drawing method. (　　)

(A) drafting.　　(B) CAD.　　(C) by instrument.　　(D) by robot.

2. A patent drawing is engineer drawing of an object that shows the relationship or order of assembly of the various parts. (　　)

(A) True.　　(B) False.

3. A3 paper size is ________ times of the A1 paper.

(A) 1/3.　　(B) 1/2.　　(C) 1/4.　　(D) 3.

4. The picture shown right side is a ________.

(A) a clothing drawing

(B) an architecture drawing

(C) a mechanical drawing

(D) a cutaway drawing

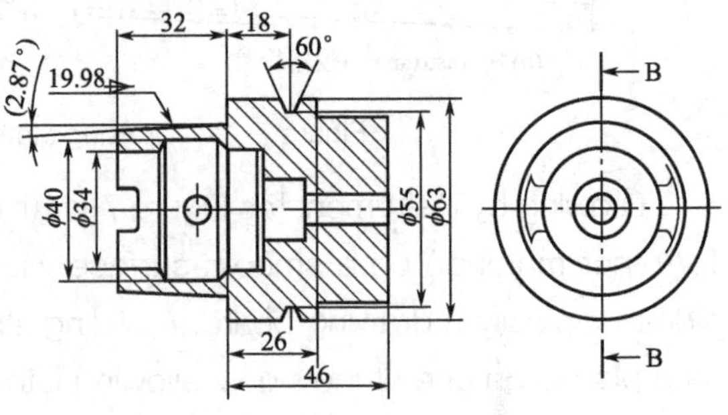

5. __________ are used for drawing simple arcs.

(A) Compasses　　(B) Triangular rule　　(C) T-square

6. Third-angle projection was an ASA standard by the __________.

(A) 1890s (B) 1950s (C) World War I (D) World War II

7. Third-angle projection is the ISO standard and is primarily used in Europe. ( )

(A) True. (B) False.

8. There are six basic views, but not all views are necessarily used. ( )

(A) True. (B) False.

9. Careless dimensioning can lead to __________ production costs or outright waste.

(A) decrease (B) reduce (C) increase

10. Mechanical drawings convey the following critical information: __________、__________、__________、__________、__________.

【Extensive Reading】/【拓展阅读】

Terms Explanation
名词解释

■ CAD/CADD

Computer-aided design (CAD), also known as computer-aided design and drafting (CADD), is the use of computer technology for the process of design and design-documentation. Computer Aided Drafting describes the process of drafting with a computer.

■ CAE

Computer-aided engineering (CAE) is the broad usage of computer software to aid in engineering tasks. CAE tools are being used, for example, to analyze the robustness and performance of components and assemblies. The term encompasses simulation, validation, and optimization of products and manufacturing tools. In the future, CAE systems will be major providers of information to help support design teams in decision making.

■ CAM

Computer-aided manufacturing (CAM) is the use of computer software to control machine tools and related machinery in the manufacturing of workpieces.

CAM is a subsequent computer-aided process after computer-aided design (CAD) and sometimes computer-aided engineering (CAE), as the model generated in CAD and verified in CAE can be input into CAM software, which then controls the machine tool.

■ CAPP

Computer-aided process planning (CAPP) is the use of computer technology to aid in the process planning of a part or product, in manufacturing. CAPP is the link between CAD and CAM in that it provides for the planning of the process to be used in producing a designed part.

EST Translation
科技英语翻译

一、翻译标准

第一，弄懂英语原文的语言，准确而完整地理解全部内容（包括思想、精神与风格），避免任意增删、曲解。

第二，汉语译文必须规范化，用词造句必须符合汉语的表达法，力求通顺易懂，不要逐词死译。为了说明翻译标准这个问题，例证如下。

【例】 The electric resistance is measured in ohms.

起初我们都会翻译成“电阻用欧姆来测量。”但是译文显得极为生硬。我们可这样译：电阻的测量单位是欧姆。

类似的句子基本上都可以套用这个格式翻译。如：…We call such a device a capacitor, or a condenser, and its ability to store electrical energy is termed capacitance. It is measured in farads. …电容的测量单位是法拉。

二、翻译过程

（一）理解阶段

在理解原文时，首先应把整篇、整章阅读一遍，对全文大意有个概念，然后逐词逐句推敲，再下笔翻译。

1. 结合上下文，推敲词意

【例】 The rate of dissociation（离解速度）was followed by placing the unit on a scale and noting the weight loss due to chlorine evolution.（离解速度是通过下述方法获得的，将该设备放置在天平上，并记录由于氯的放出而减少的重量。）

scale有多种意思：scale 天平盘，scales 天平/磅秤，scale标度/刻度，scale（音）音阶。这是涉及化学的句子，在这里应翻译成天平。

【例】 Various speeds may be obtained by the use of large and small pulley.（利用大小皮带轮，可以获得不同的速度。）

补充知识：fixed pulley 定滑轮，movable pulley 动滑轮，block and tackle 滑轮。

2. 辨明语法，弄清关系

【例】 As friction manifests itself as a resistance that opposes motion, it is usually considered as a nuisance.（摩擦是运动的阻力，因此常被人看作是讨厌的东西。）

【例】 Intense light and heat in the open contrasted with the coolness of shaded avenues and the interiors of buildings.（露天场所的强烈光线和酷热同林荫道和建筑内部的凉爽形成对比。）

（二）表达阶段

表达就是选择恰当的汉语，把已经理解的原作内容重新叙述出来，有直译和意译两种方法。一般来说，意译比直译难度大。意译在头脑中经过再加工的成分更多，是一个重新将信息整合的过程。

1. 直译

【例】 Industrial regions of the world suffer much more acidic fall-out than they did before the industrial revolution.（现在，世界上的工业地区比工业革命前遭受到更多酸性沉降物。）

2. 意译

【例】 Mankind has always reverenced what Tennyson call "the useful trouble of the rain".（人们一直很推崇特尼森所说的话：雨既有用又带来麻烦。）

【例】 In fact, it may be said that anything that is not animal or vegetable is mineral.（任何东西只要既不是动物又不是植物便是矿物。）

三、词义处理

根据词的搭配确定词义。

【例】 wet

Wet paint!（油漆未干！）

He was wet to the skin.（他全身湿透了。）

All that time he was still wet behind the ears.（那时他还乳臭未干。）

If you think I am for him, you are all wet.（如果你认为我支持他，你就大错特错了。）

【例】 work

The machine works properly.（这台机器运转正常。）

The mine has long been worked.（这个矿场已经开采很久了。）

The threads of the screw work hard.（这个螺纹太涩了。）

四、科技英语中部分否定句的汉译

在英语的否定结构中，由于习惯用法问题，其中部分否定句所表示的意思是不能按字面顺序译成汉语的，因此，翻译时要特别注意。英语中含有全体意义的代词和副词如all、every、both、always、altogether、entirely等统称为总括词。它们用于否定结构时不是表示全部否定，而只表示其中的一部分被否定。因此，汉译时不能译作“一切……都不”，而应译为“并非一切……都是的”或“一切……不都是”。

【例】All of the heat supplied to the engine is not converted into useful work.（并非供给热机的所有热量都被转变为有用的功。错译：所有供给热机的热量都没有被转变为有用的功。）

【例】Every one cannot do these tests.（并非人人都能做这些试验。错译：每个人都的不能做这些试验。）

【例】Both instruments are not precise.（两台仪器并不都是精密的。错译：两台仪器都不是精密的。）

【例】This plant does not always make such machine tools.（这个工厂并不总是制造这样的机床。错译：这个工厂总是不制造这样的机床。）

但是当出现（总括词 + 肯定式谓语 + 含否定意义的单词……）这种结构时，则是表示全部否定。

【例】All germs are invisible to the naked eye.（一切细菌都是肉眼看不见的。）

【例】Every design made by her is impossible of execution.（她所做的一切设计都是不能执行的。）

【例】Both data are incomplete.（两个数据都不完整。）

【例】In practice, error sometimes always seems unavoidable.（在实践中，差错有时似乎总是不可避免的。）

五、定语从句的汉译

（1）逆序合译法　只要是比较短的，或者虽然较长，但汉译后放在被修饰语之前仍然很通顺的，一般的就放在被修饰语之前。

【例】The speed of wave is the distance it advances per unit time.（波速是波在单位时间内前进的距离）。

【例】The light wave that has bounced off the reflecting surface is called the reflected ray.（从反射表面跳回的光波称为反射线。）

【例】Stainless steel, hick is very popular for its resistance to rusting, contains large percentage of chromium.（具有突出防锈性能的不锈钢含铬的百分比很高。）

（2）顺序分译法　定语从句较长，或者虽然不长，但汉译时放在被修饰语之前实在不通顺的就后置，作为词组或分句。

【例】 Each kind of atom seems to have a definite number of “hands” that it can use to hold on to others.

每一种原子似乎都有一定数目的“手”，用来抓牢其他原子。(顺序分译法)

每一种原子似乎都有一定数目用于抓牢其他原子的“手”。(逆序合译法)

这句限制性定语从句虽然不长，但用顺序分译法译出的译文要比用逆序合译法更为通顺。

【例】 Let AB in the figure above represent an inclined plane the surface of which is smooth and unbending.

设上图中AB代表一个倾斜平面，其表面光滑不弯。(顺序分译法)

设上图中AB代表一个其表面为光滑不弯的倾斜平面。(逆序合译法)

上面两种译法，也是用顺序分译法比用逆序合译法更为通顺简明。

（3）定语从句较长，与主句关联又不紧密，汉译时就作为独立句放在主句之后。这种译法仍然是顺序分译法。

【例】 Such a slow compression carries the gas through a series of states, each of which is very nearly an equilibrium state and it is called a quasi-static or a “nearly static” process.（这样的缓慢压缩能使这种气体经历一系列的状态，但各状态都很接近于平衡状态，所以叫做准静态过程或“近似稳定”过程。）

【例】 Friction wears away metal in the moving parts, which shortens their working life.（运动部件间的摩擦力使金属磨损，这就缩短了运动部件的使用寿命。）

（4）汉译there + be句型中的限制性定语从句时，往往可以把主句中的主语和定语从句溶合在一起，译成一个独立的句子。这种译法叫做溶合法，也叫拆译法。

【例】 There are bacteria that help plants grow, others that get rid of dead animals and plants by making them decay, and some that live in soil and make it better for growing crops.（有些细菌能帮助植物生长，另一些细菌则通过腐蚀来消除死去的动物和植物，还有一些细菌则生活在土壤里，使土壤变得对种植庄稼更有好处。）

【例】 There is a one-seated which you could learn to drive in fifty minutes.（有一种单座式汽车，50min就能让你学会驾驶。）

Chapter 8

Products Test
产品检测

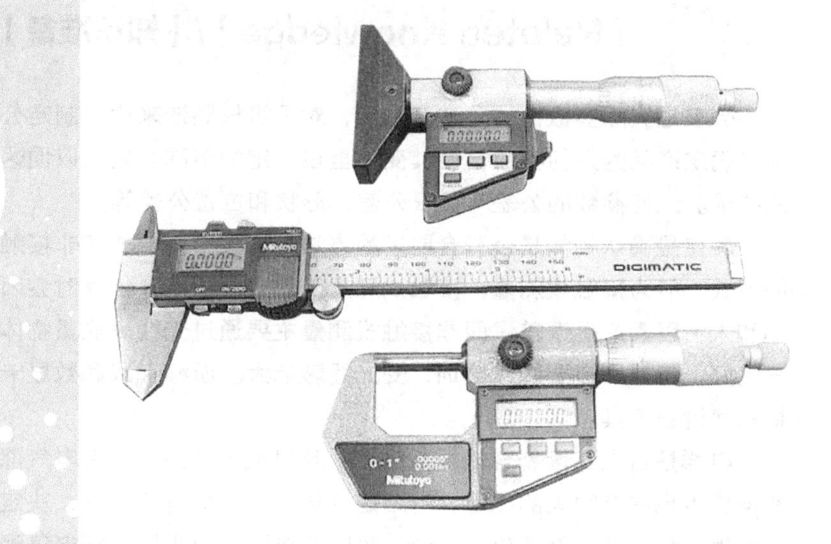

【Content Description】/【内容描述】

为了保证各种产品正常工作，比如电梯能安全运行，各国相关部门制定了一系列标准和法规。为了确认整机或零部件的尺寸或功能是否达到这些标准，就需要通过检测或测试进行确认；检测也用于工厂自检或确认加工件是否达到设计要求；进行故障诊断和排除时，也需要进行测量和检验以确认问题所在。

本章包括三节，公差和质量控制，尺寸测量和电气测试。第一节主要内容包括公差概念及用途，极限公差与形位公差，公差带与配合，质量控制；第二节主要介绍测量工具及其使用，如游标卡尺、厚薄规；第三节电气测试部分主要介绍万用表的使用。通过以上内容的学习，应建立检测的概念和意识，掌握常见的测量方法和相关术语。

【Related Knowledge】/【知识准备】

公差是实际参数值的允许变动量，对于机械制造来说，制定公差的目的就是为了确定产品的几何参数，使其变动量在一定的范围之内，以便达到互换或配合的要求。几何参数的公差有极限公差、形状和位置公差等。

测量是确认物体是否符合要求的方法，可分为接触式和非接触式测量。传统的方式一般为接触式测量，接触式测量简单方便快捷，但有时会因为夹紧力的大小和人为因素产生误差；而非接触式测量主要通过光线来测量物体，精度高、统一性好，但操作需要经过培训，受光线影响大。游标卡尺和数显卡尺是最常用的接触式测量工具。

电梯是机电高度整合的复杂系统，除机械部分外，还有电气部分。电气检测在电梯中也占有很大的比重，电路是否导通、接触是否良好，决定了电梯运行的安全性、稳定性和舒适感。另外，故障的排除，也需要进行电气测量来诊断。万用表是电气测量中最常用的设备。

【Section Implement】/【章节内容】

Section 1 Tolerance and Quality Control 公差和质量控制

A test of the completed product is often nothing more than a contractual requirement that must be performed before the customer accepts the products. It allows for gathering data that support the design theory of the product, for interpretations to be made for further improvements in design so that future products will be better than present ones, and for evaluation of design evolution toward better performances costs. In addition, it is a means of verifying design, since not all design parameters can be fully calculated or predicted.

Product test engineers work closely with design engineers to provide useful data for testing. They must also work in close harmony with engineers all other phases of manufacturing. Not infrequently, product testing will turn up deficiencies in design that requires major revisions in manufacturing processes. This particularly true if the company produces many prototypes and has short production runs. Therefore, manufacturing engineers are as interested in product test's results as are design engineers.

For complex products, product test becomes a very important part of the total process control function. It gives the company a high degree of confidence that the product will perform as the customer expects it to, and this is a valuable marketing tool as it helps to establish the proper reputation with the customers.

▪ Deviation Tolerances 极限公差

Control of dimensions is necessary in order to ensure assembly and interchangeability of components. Tolerances are specified on critical dimensions that affect clearances and interferences fit. One method of specifying tolerances is deviation tolerances. This dictates how far away from the nominal dimension, the actual measurement is allowed to be, so a dimension could be stated as Φ40.0+/−0.3mm. This means that the dimension should be machined between Φ39.7mm and Φ40.3mm.

If the variation can vary either side of the nominal dimension, the tolerance is called a bilateral tolerance, such as $\Phi 40 \pm 0.03$ or $\Phi 40^{+0.4}_{-0.1}$. For a unilateral tolerance, one tolerance is zero, such as $\Phi 40^{0.5}_{0}$ or $\Phi 40^{0}_{-0.5}$. An example can be seen in figure8−1.

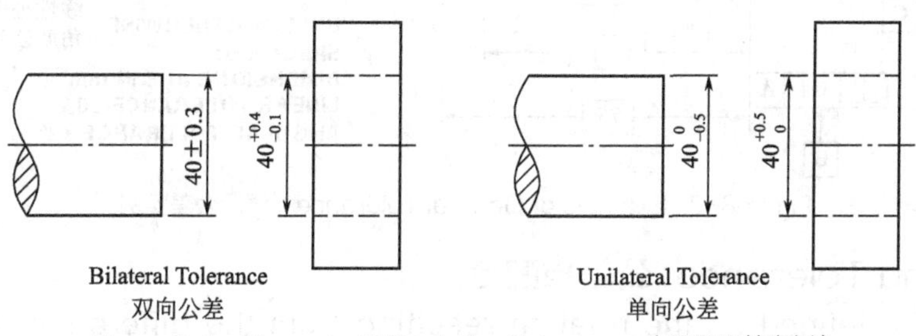

Figure8−1 Example of Deviation Tolerances（极限公差实例）

Most organizations have general tolerances (as Figure 8–2 shows) that apply to dimensions when an explicit dimension is not specified on a drawing. For machined dimensions a general tolerance maybe +/–0.5mm. So a dimension specified as 15.0mm may range between 14.5mm and 15.5mm. Other general tolerances can be applied to features such as angles, drilled and punched holes, castings, forgings, weld beads and fillets. General tolerance has another name, default tolerance.

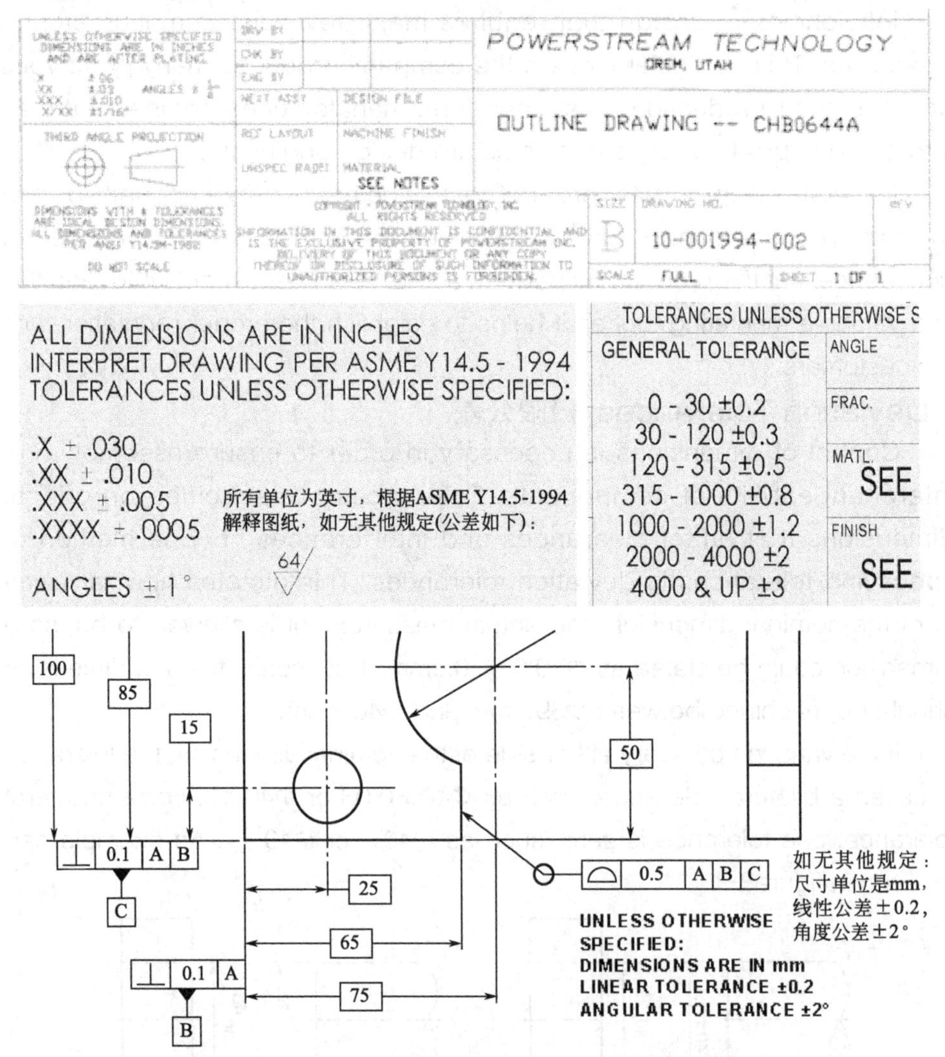

Figure8–2 Example of General Tolerance（通用公差实例）

▪ Fit and Tolerance 公差与配合

Fit is defined as the relation resulting from the difference between the sizes of two mating parts. And tolerance is defined as the difference between the maximum limit and the minimum limit. The tolerance is also equal to algebraic difference between the upper and lower deviations. Depending upon the actual limits of the hole or shaft, a fit may be classified as follows (as Figure 8-3 shows) :

A. Clearance Fit is a fit that always provides a clearance between the mating parts. In this case, the tolerance zone of the hole is entirely above that of the shaft.

B. Transition Fit is a fit that may provide either a clearance or interference between the mating parts depends on the actual dimensions of the finished parts.

C. Interference Fit is a fit that always provides interference between the mating parts. Here, the tolerance zone of the hole is entirely below that of the shaft.

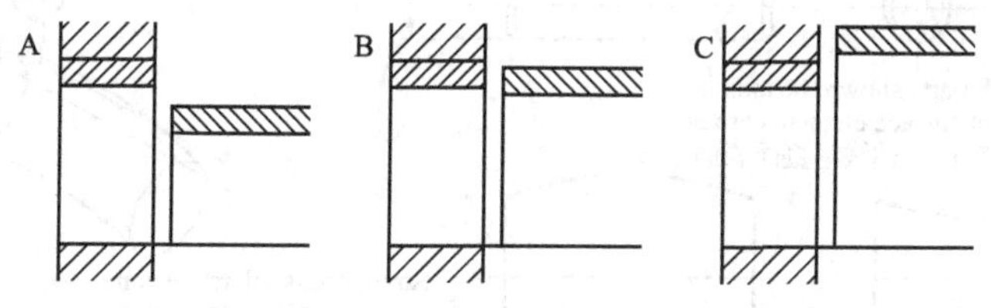

Figure8-3　Tolerance Zone and Fit Types（公差带与配合类型）

In this situation, a tolerance is designated by a letter (in some case, two letters) symbol and a numerical symbol, such as 20H7 or 20g6. Capital letters are used for holes and small letters for shafts. The letter symbol indicates the position of the zone of tolerance in relation to the zero line representing the basic size. The numerical symbol represents the value of this zone of tolerance and is called the grade or quality of tolerance. Both the position and the grade of tolerance are functions of the basic size.

▪ Geometric Tolerance 几何公差

Geometric Tolerance (as Figure 8-4 shows), also called geometric dimensioning and tolerancing (GD&T), is a system for defining and communicating engineering tolerances. It uses a symbolic language on engineering drawings and computer-generated three-dimensional solid models for explicitly describing nominal geometry and its allowable variation. It tells the manufacturing staff and machines what degree of accuracy and precision is needed on each facet of the part (Example see Figure 8-5 shows).

Figure8-4　Geometric Tolerance（几何公差）

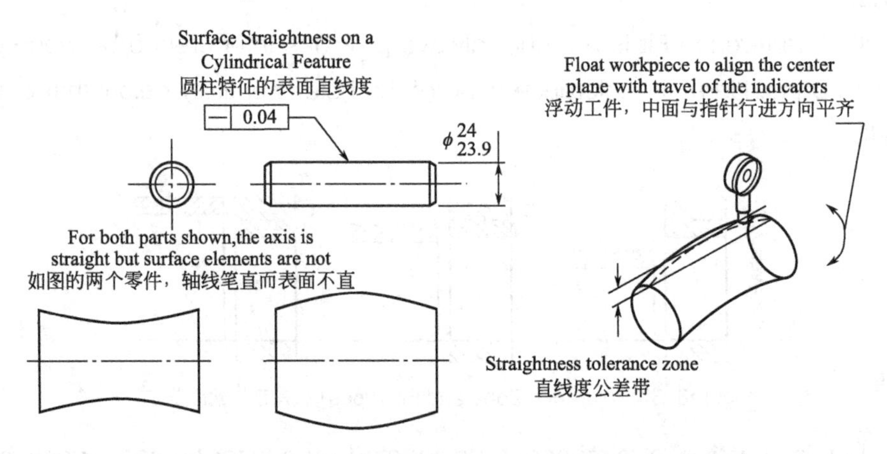

Figure8-5　Example of Straightness Error（直线度误差实例）

▪ Quality Control 质量控制

Quality control has traditionally been the liaison between manufacturing and design. This function interprets design's specifications for manufacturing and develops the quality pan to be integrated into manufacturing engineering's methods and planning instructions to operations. Quality control is also responsible for recommending to management what level of manufacturing losses (cost of mistakes in producing the product) can be tolerated. This is based on the complexity of the product design; specifically the degree of preciseness necessary in tolerances. Quality control traditionally monitors manufacturing losses by setting a negative budget that is not to be exceeded, and establishes routines for measurement and corrective action.

Within the past decade or two, quality control has become increasingly involved with marketing and customers in establishing documentation systems to ensure guaranteed levels of product quality. This new role has led to the new title quality assurance, to differentiate it from traditional in-house quality control.

Quality assurance strives through documentation of performance and characteristics at each stage of manufacture to ensure that the product will perform at the intended level. Whereas quality control is involved directly with manufacturing operations, quality assurance is involved with the customer support responsibilities generally found within the marketing function. Many industrial organizations have chose to establish an independent quality assurance sub-function within the manufacturing function and have placed the technical responsibilities of quality control, namely process control, within the manufacturing engineering organization.

Notes and Expressions

1. short production run 小批生产，短期生产
2. bilateral tolerance 双向公差；双边公差
3. unilateral tolerance 单向公差；单边公差
4. liaison [lje'zuŋ] *n.* 联系；联络；联系人
5. dictate [dik'teit; 'dikteit] *vt.* 口授；命令；指定；决定
6. general tolerance 通用公差；未注公差；一般公差
7. algebraic [ˌældʒi'breiik] *adj.* 代数(学)的；代数上的

Section 2 Dimension Measurement 尺寸测量

▪ Caliper 卡尺

The vernier, dial, and digital calipers give a direct reading of the distance measured to high accuracy. They are functionally identical, with different ways of reading the result. These calipers comprise a calibrated scale with a fixed jaw, and another jaw with a pointer, that slides along the scale.

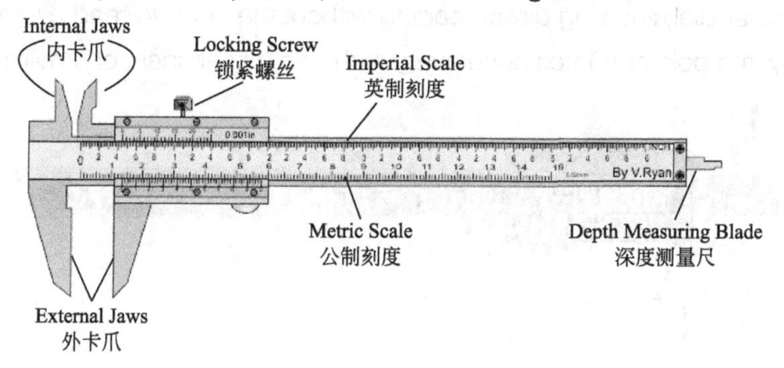

Figure 8-6 Vernier Caliper(游标卡尺)

Vernier, dial, and digital calipers can measure internal dimensions using the uppermost jaws in the picture, external dimensions using the pictured lower jaws, and in many cases depth by the use of a probe that is attached to the movable part and slides along the centre of the body. This probe is slender and can get into deep grooves that may prove difficult for other measuring tools.

The distance between the jaws is then read in different ways for the three types. The simplest method is to read the position of the pointer directly on the scale. When the pointer is between two markings, the addition of a vernier scale allows more accurate interpolation, this is the vernier caliper.

The vernier (as Figure 8-6 shows) scales may include metric measurements on the lower part of the scale and inch measurements on the upper, or vice versa. Vernier calipers commonly used in industry provide a precision to a hundredth of a millimeter or 10 micrometers, or one thousandth of an inch. They are available in sizes that can measure up to 72 in (1800 mm). The slide of a vernier caliper can usually be locked at a setting using a small screw, this allows simple go/no-go checks of part sizes.

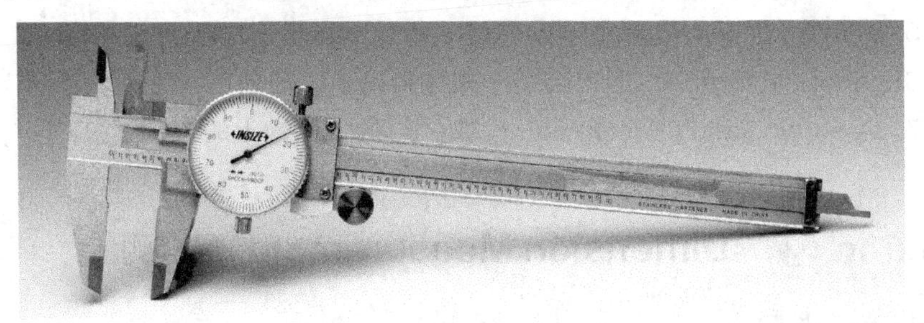

Figure 8-7 Dial Caliper（转盘卡尺）

Instead of using a vernier mechanism, which requires some practice to use, the dial caliper (as Figure 8-7 shows) reads the final fraction of a millimeter or inch on a simple dial. In this instrument, a small but precise gear rack drives a pointer on a circular dial, allowing direct reading without the need to read a vernier scale. Typically, the pointer rotates once every inch, tenth of an inch, or 1 millimeter.

Figure 8-8 Digital Caliper（数显卡尺）

A refinement now popular is the replacement of the analog dial with an electronic digital display on which the reading is displayed as a single value. Some digital calipers (as Figure 8-8 shows) can be switched between centimeters or millimeters, and inches. All provide for zeroing the display at any point along the slide, allowing the same sort of differential measurements as with the dial caliper. The digital interface significantly decreases the time to make and record a series of measurements, and it also improves the reliability of the records.

Ordinary 6-in/150-mm digital calipers are made of stainless steel, have a rated accuracy of 0.02mm, the same technology is used to make longer 8-in and 12-in calipers.

Digital calipers contain a capacitive linear encoder. A pattern of bars is etched directly on the printed circuit board in the slider. Under the scale of the caliper another printed circuit board also contains an etched pattern of lines. The combination of these printed circuit boards forms two variable capacitors. The two capacitances are out of phase. As the slider moves the capacitance changes in a linear fashion and in a repeating pattern. The circuitry built into the slider counts the bars as the slider moves and does a linear interpolation based on the magnitudes of the capacitors to find the precise position of the slider.

▪ Micrometer Caliper 千分尺

A caliper using a calibrated screw for measurement, rather than a slide, is called a micrometer caliper (as Figure 8-9 shows) or, more often, simply a micrometer. Micrometers typically have a specified temperature at which the measurement is correct (often 20℃ or 68℉, which is generally considered “room temperature”).

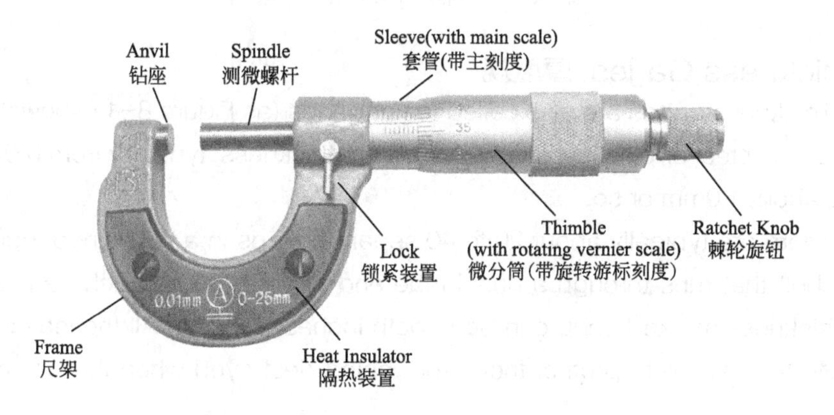

Figure 8-9 Micrometer Caliper（千分尺）

▪ Measuring Tape 卷尺

A tape measure or measuring tape (as Figure 8-10 shows) is a flexible form of ruler and it is a common measuring tool. It consists of a ribbon (cloth, plastic, fiber glass, or metal strip) with linear-measurement markings. Its flexibility allows for a measure of great length to be easily carried in pocket or toolkit and permits one to measure around curves or corners. Surveyors use tape measures in lengths of over 100 m (300+ ft).

Today, measuring tapes made for sewing are made of fiberglass, which does not tear or stretch as easily. Measuring tapes designed for carpentry or construction often use a stiff, curved metallic ribbon that can remain stiff and straight when extended, but retracts into a coil for convenient storage. This type of tape measure will have a floating tang on the end to aid measuring. The tang will float a distance equal to its thickness, to provide both inside and outside measurements that are accurate.

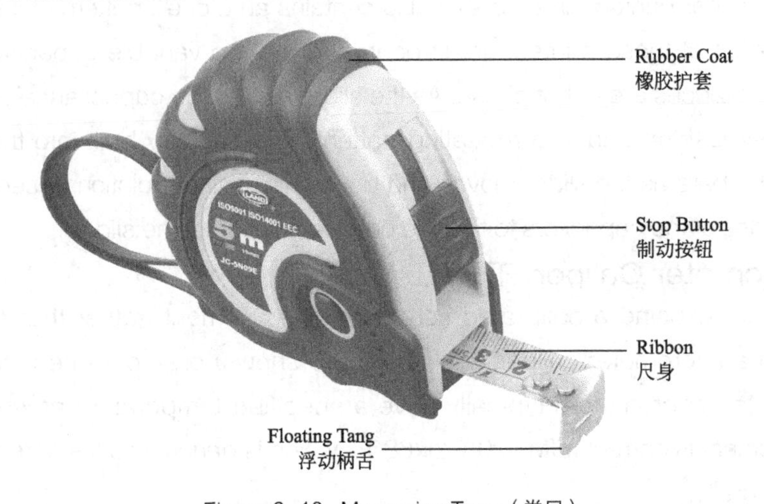

Figure 8-10 Measuring Tape（卷尺）

▪ Thickness Gages 厚度规

Thickness gages are also called feeler gages (as Figure 8-11 shows), are sets of thin steel or plastical strips of accurate thickness, typically from 0.02mm up to about 1.0mm or so.

There are typically about 20 to 40 separate strips in a set, joined together by a bolt that runs through a hole in the end of each gage. Each gauge has the thickness marked on it, can be in both inches and mm. All the leaves fold up into the handle to protect them from being destroyed when they're not in use.

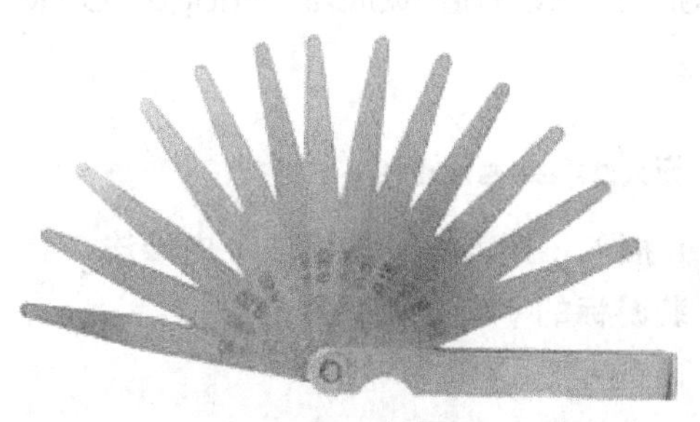
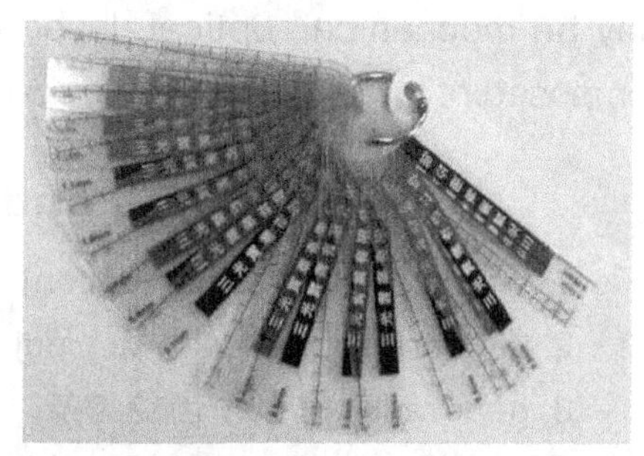

Figure 8-11 Feeler Gages（厚薄规）

▪ Height Gauge 高度计

A height gauge (as Figure 8-12 shows) is a measuring device used either for determining the height of something, or for repetitious marking of items to be worked on. The former type of height gauge can be used in doctor's surgeries to find the height of people. The latter measuring tools are used in metalworking or metrology to either set or measure vertical distances; the pointer is sharpened to allow it to act as a scriber and assist in marking out work pieces.

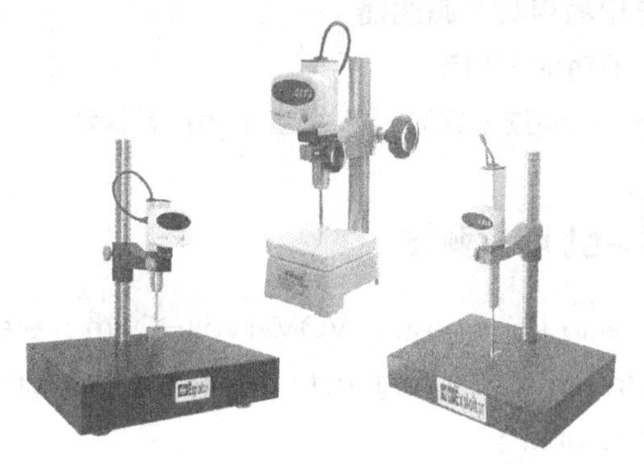

Figure 8-12 Height Gauge（高度计）

They may also be used to measure the height of an object by using the underside of the scriber as the datum. The datum may be permanently fixed or the height gauge may have provision to adjust the scale, this is done by sliding the scale vertically along the body of the height gauge by turning a fine feed screw at the top of the gauge; then with the scriber set to the same level as the base, the scale can be matched to it.

▪ CMM 三坐标测量仪

A coordinate measuring machine (CMM) is a device for measuring the physical geometrical characteristics of an object. This machine may be manually controlled by an operator or it may be computer controlled. Measurements are realized by a probe attached to the machine. Probes may be mechanical, optical, laser, or others. The typical "bridge" CMM is composed of three axes, an X, Y and Z.

Notes and Expressions

1. interpolation [in,tə:pəu'leiʃən] *n.* 插入；插入物；插值法；内推法
2. vernier ['və:niə] *n.* 游标 adj. 装游标的；有微调装置的
3. gear rack 齿条；齿轮齿条
4. ratchet ['rætʃit] *n.* 棘轮；棘齿 *vt.* 安装棘轮于
5. sleeve [sli:v] *n.* 套筒，袖子 *vt.* 给……装袖子/套筒
6. knob [nɔb] *n.* 把手；瘤；球形突出物 *vi.* 鼓起
7. metrology [mi'trɔlədʒi] *n.* 度量衡；度量衡学
8. orthogonal [ɔ:'θɔgənəl] *adj.* [数] 正交的；直角的 *n.* 正交直线
9. regression algorithms ['ælgəriðəm] *n.* 回归算法
10. ribbon ['ribən] *n.* 缎带；丝带；带状物；钢卷尺；带锯；(钟表的)发条
11. scriber ['skraibə] *n.* 描绘标记的用具；画线器
12. manually ['mænjuəli] *adv.* 手动地；用手
13. probe [prəub] *n.* 探针；调查 *vi.* 调查；探测 *vt.* 探查；用探针探测

Section 3 Electrical Test 电气测试

A multimeter or a multitester, also known as a VOM (Volt-Ohm meter) (as Figure 8-13 shows), is an electronic measuring instrument that combines several measurement functions in one unit.

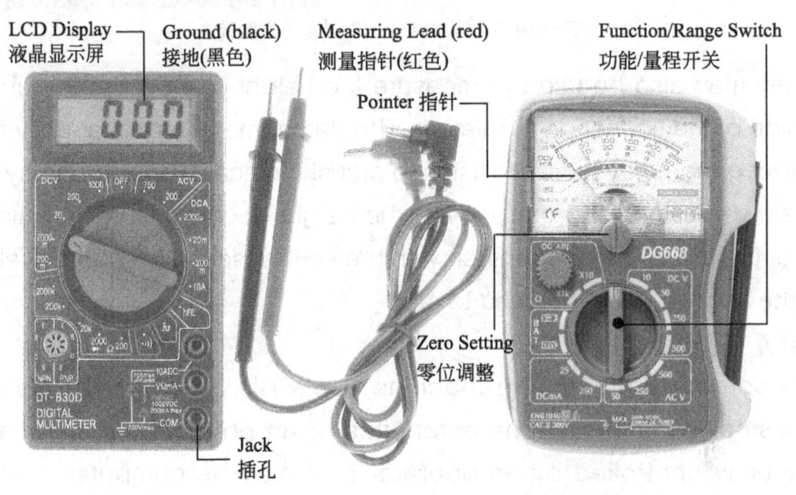

Figure 8-13 Digital and Analog Multimeter（数字和模拟万用表）

By operating a multi-position switch on the meter, it can be quickly and easily set to be a voltmeter, ammeter and ohmmeter. They have several settings (called ranges) for each type of meter and the choice of AC or DC. Some multimeters have additional features such as transistor testing and ranges for measuring capacitance and frequency. A multimeter can be a useful hand-held device used to troubleshoot electrical problems in a wide array of industrial and household devices

There are two basic types of multimers, digital and analog. Analog multimeters have a needle and the digital has a LED display.

▪ Analog Multimeter 模拟万用表

Resolution of analog multimeters is limited by the width of the scale pointer, vibration of the pointer, the accuracy of printing of scales, zero calibration, number of ranges, and errors due to non-horizontal use of the mechanical display.

Resistance measurements, in particular, are of low precision due to the resistance measurement circuit. Inexpensive analog meters may have only a single resistance scale, seriously restricting the range of precise measurements. Typically an analog meter will have a panel adjustment to set the zero-ohms calibration of the meter, to compensate for the varying voltage of the meter battery.

▪ Digital Multimeter 数字万用表

A digital multimeter is one of the most useful and helpful tools in our house and during the job. It is important to own a good type and learn how to utilize it in a proper manner. Today, contemporary digital multimeters are designed to be rugged and easy to operate. A good multimeter will include a rugged plastic case and huge, easy to use selector knobs.

The top part is the digital read out monitor. This is something you should thoroughly check out before you purchase one. Be sure the screen is huge enough to read it and make sure you notice the readout in sunlight. Possibility is you will be utilizing this instrument outside in direct sunlight. Be certain the function switch is large and simple to operate. The majority of function switches have eight positions.

Remember to shut off your device before keeping it back in your toolbox. It is should be done for the purpose to save battery the next time you will have to use it.

▪ The Use of Multimeter 万用表的使用

Multimeters are easily damaged by careless use, so please take these precautions:

1. Select a range with a maximum greater than you expect the reading to be.

2. Connect the meter, making sure that the leads are the correct way around. Digital meters can be safely connect in reverse, but an analogue meter may be damaged.

3. If the reading goes off the scale: immediately disconnect and select a higher range.

4. Always disconnect the multimeter before adjust the range switch

5. Always check the setting of the range switch before you connect to a circuit.

6. Never leave a multimeter set to a current range except when actually taking a reading. The greatest risk of damage is on the current ranges because the meter has a low resistance.

When testing circuits you often need to find the voltages at various points, for example the voltage at pin2 of a 555 timer chip shown in figure8-14. This can seem confusing—where should you connect the second multimeter lead?

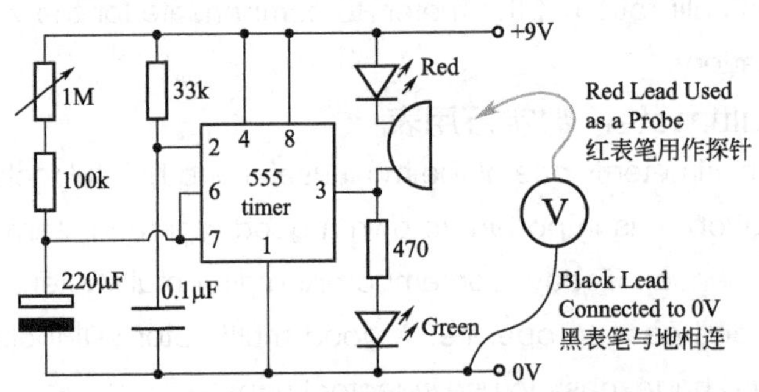

Figure 8-14 Measuring Voltage at a point（测量某一点电压）

Connect the black lead to 0V, normally the negative terminal of the battery or power supply. Then connect the read lead to the point where you need to measure the voltage. The black lead can be left permanently connected to 0V, while you use the read lead as a probe to measure voltages at various points.

Another useful function of the DMM is the ohmmeter. An ohmmeter measures the electrical resistance of a circuit. If there is no resistance in a circuit, the ohmmeter will read 0; if there is an open in a circuit, it will read infinite.

Notes and Expressions

1. microammeter [ˌmaikrəu'æmitə] *n.* [电] 微安计

2. All these can be quickly and easily set by operating a multi-position switch. 所有这些都可以通过操作切换开关快捷而方便地实现。by是介词，后面要跟名词、动名词或动词现在分词。

3. Select a range with a maximum greater than you expect the reading to be. 根据预期的读数范围选择大一级的量程。句中reading意为读数。

4. When testing circuits you often need to find the voltages at various points, for example the voltage at pin2 of a 555 timer chip. 当进行电路测量时，常常需要知道某一点的电压。在从句和主句有的主语相同时，可以将从句主语省略，并将谓语改为动词现在分词。从句when testing circuits的主语为you，与主句you often need to find the voltages…中的主语you相同，故省略。

【Self Evaluation】/【自我评价】

1. Digital multimeter is more convenient than analogue Multimeter. (　　)

(A) True.　　(B) False.

2. Multimeter can be used to test both DC and AC voltage. (　　)

(A) True.　　(B) False.

3. Straightness of motion of a linear bearing is ________.

(A) deviation tolerance　　(B) geometric tolerance

(C) general tolerance　　(D) bilateral tolerance

4. __________ always provides a gap between the mating parts.

(A) Interference fit　　(B) Transition fit

(C) Clearance fit　　(D) None fit

5. The __(1or2) capacitance(s) of digital caliper are out of phase.

(A) 1　　(B) 2　　(C) 3　　(D) 4

6. A micrometer using __________ for measurement.

(A) a slide　　(B) two capacitances

(C) gear rack and circular dial　　(D) a calibrated screw

7. The material of feeler gages can be ___________.

(A) steel or glass　　(B) steel or plastics

(C) polymer or steel (D) polymer or glass

8. Mulitmeter can be a hand-held device useful for basic fault finding and it can be used to troubleshoot electrical problems. ( )

(A) True. (B) False.

9. There is a capital letter "H" in 20H7, so it is used for ________.

(A) shafts (B) holes

(C) both shafts and holes (D) either holes or shafts

10. Tolerance zone of hole is ________ that of the shaft for interference fit.

(A) entirely above (B) parallel

(C) overlapping (D) entirely below

11. Vernier caliper can be used to measure __________、__________ and __________.

【Extensive Reading】/【拓展阅读】

Elevator Test
电梯测试

▪ Elevator Test Tower 电梯测试塔

An Elevator test tower is a structure usually 100 to 140 meters tall that is designed to evaluate the stress and fatigue limits of specific elevator cars in a controlled environment. Tests are also carried out in the test tower to insure reliability and safety in current elevator designs and address any failures that may arise. Examples of an elevator test tower are the Express Lift Tower in Northampton, England, and the Shibayama Test Tower in Shibayama, Japan.

Mitsubishi has erected a tall, skinny, hollow tower filled with elevator shafts for testing high-speed lifts. The 173m-high (567ft) structure is called Solae and dominates the skyline of Inazawa City. The 5bn-yen (equals $50m) project will allow Mitsubishi to test new drives, gears and other lift systems.

▪ Reliability Test 可靠性测试

The purpose of reliability testing is to discover potential problems with the design as early as possible and, ultimately, provide confidence that the system meets its reliability requirements.

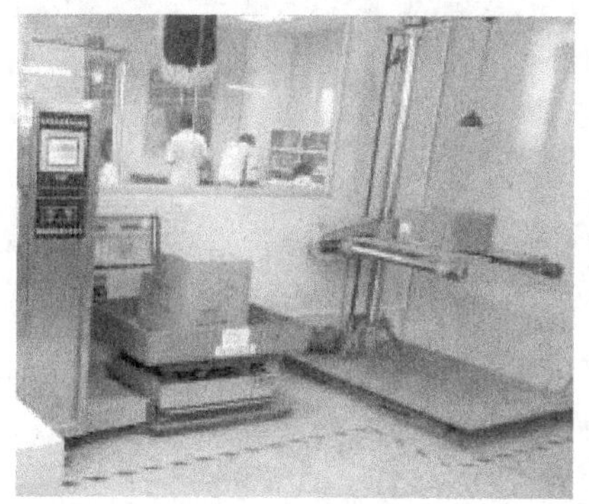

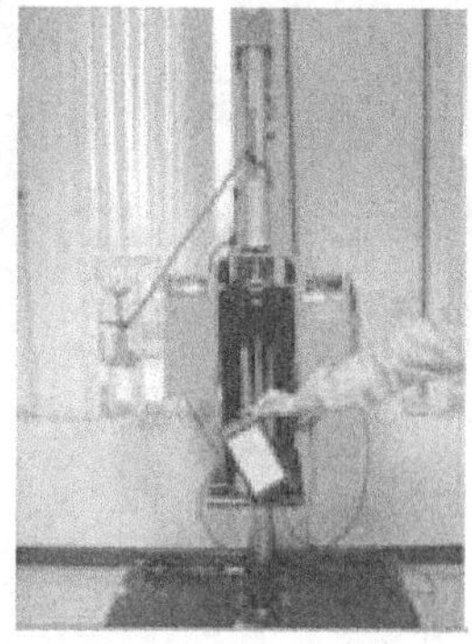

It is not always feasible to test all system requirements. Some systems are prohibitively expensive to test; some failure modes may take years to observe; some complex interactions result in a huge number of possible test cases; and some tests require the use of limited test ranges or other resources. In such cases, different approaches to testing can be used, such as accelerated life testing, design of experiments, and simulations.

▪ Accelerated Testing 加速测试

The purpose of accelerated life testing is to induce field failure in the laboratory at a much faster rate by providing a harsher, but nonetheless representative, environment. In such a test the product is expected to fail in the lab just as it would have failed in the field—but in much less time.

The main objective of an accelerated test is either of the following: To discover failure modes and To predict the normal field life from the high stress lab life.

Chapter 9

Electrical System Overview 电气系统简介

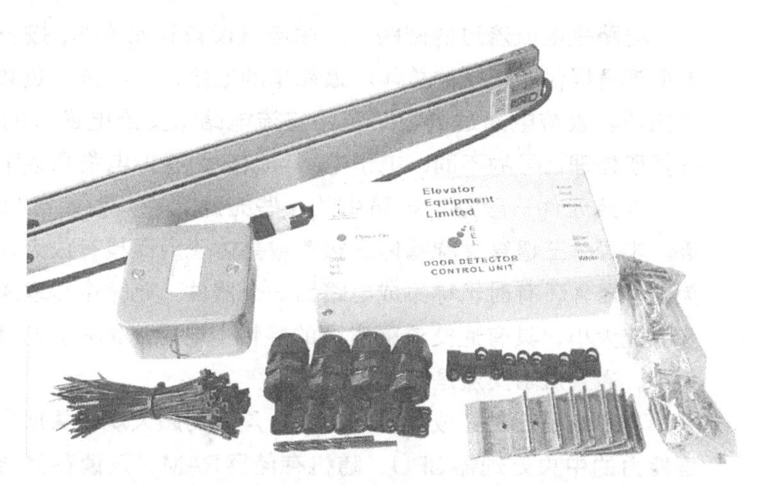

【Content Description】/【内容描述】

电气系统是电梯中的重要部分，负责电梯升降指令的发出和接受，电梯运行状态的监控，紧急状态下的保护等诸多功能。系统由多个电路组合而成，不同的电路由功能各异的电气元件和导线构成。

本章包括两节，电路、电器元件，单片机和可编程控制器等电气知识。通过对本章内容的学习，包括通过对话练习，掌握常见的英文单词，能听懂电路及其结构的描述，进行基本的工作交流和讨论。

【Related Knowledge】/【知识准备】

电路是电流流过的回路，是由电气设备和元器件，按一定方式连接起来，为电荷流通提供了路径的总体。最简单的电路，由电源、负载、导线、开关等元器件组成。直流电通过的电路称为直流电路；交流电通过的电路称为交流电路。根据所处理信号的不同，电子电路可以分为模拟电路和数字电路。

电路元件一般是指电路中的一些无源元件，譬如电阻器、电容器、电感器等。半导体三极管又称晶体三极管或晶体管。而具有放大功能的晶体管等往往称为电子器件，有时也称集成电路为一个器件。电路中C取决于电压随时间变化率和电流大小，具有通交流隔直流的特性；电感L取决于电流随时间变化率和电压大小，通直流阻交流是其特性。

单片机是一种集成的电路芯片，是采用超大规模集成电路技术把具有数据处理能力的中央处理器CPU、随机存储器RAM、只读存储器ROM、多种I/O口和中断系统、定时器/计时器等功能集成到一块硅片上构成的一个小而完善的计算机系统。可编程逻辑控制器是一种数字运算操作的电子系统，专为工业应用而设计。它采用一类可编程的存储器，用于其内部存储程序，执行逻辑运算，顺序控制，定时，计数与算术操作等面向用户的指令，并通过数字或模拟输入/输出控制各种类型的机械或生产过程。

【Section Implement】/【章节内容】

Section 1 Electrical Circuit 电路

Read the following passages and try to describe circuit in your own words with English.

▪ What's Electrical Circuit? 什么是电路

The picture shows the basic type of electrical circuit, and the electrical circuit can be described in the form of a diagram as figure9-1. It consists of a source of electrical energy, some sort of load to make use of that energy, and electrical conductors connecting the source and the load.

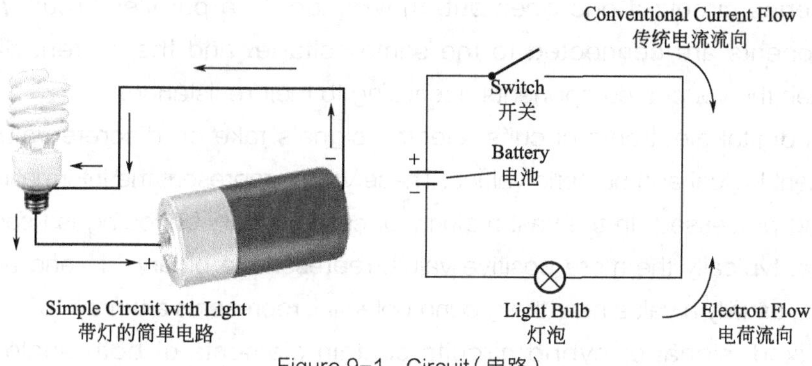

Figure 9-1 Circuit(电路)

As long as there is an unbroken connection from source to load and back again as shown here, electrons will be pushed from the negative terminal of the source, through the load, and then back to the positive terminal of the source. The arrows show the direction of electron current flow through this circuit. Because the electrons are always moving in the same direction through the circuit, their motion is known as direct current (DC).

The source can be any source of electrical energy. In practice, there are three general possibilities: it can be a battery, an electrical generator, or some sort of electronic power supply. The load is any device or circuit powered by electricity. It consumes the electricity from a circuit and converts it into work — heat and light. The load can be as simple as a light bulb or as complex as a modern high-speed computer.

▪ Types of Circuits 电路种类

Three types of electrical circuits exist: the series circuit, the parallel circuit, and the series-parallel circuit. A series circuit is the simplest because it has only one possible path that the electrical current may flow; if the electrical circuit is broken, none of the load devices will work. The difference with parallel circuits is that they contain more than one path for electricity to flow, so if one of the paths is broken, the other paths will continue to work. A series-parallel circuit, however, is a combination of the first two: it attaches some of the loads to a series circuit and others to parallel circuits.

Analog electronic circuits are those in which current or voltage may vary continuously with time to correspond to the information being represented. Analog circuitry is constructed from two fundamental building blocks: series and parallel circuits. In a series circuit, the same current passes through a series of components. A string of Christmas lights is a good example of a series circuit: if one goes out, they all do. In a parallel circuit, all the components are connected to the same voltage, and the current divides between the various components according to their resistance.

In digital electronic circuits, electric signals take on discrete values, to represent logical and numeric values. These values represent the information that is being processed. In the vast majority of cases, binary encoding is used: one voltage, typically the more positive value, represents a binary '1' and another voltage, usually a value near the ground potential, represents a binary '0'.

Mixed-signal or hybrid circuits contain elements of both analog and digital circuits. Examples include comparators, timers, ADCs (analog-to-digital converters), and DACs (digital-to-analog converters). Most modern radio and communications circuitry uses mixed signal circuits. For example, in a receiver, analog circuitry is used to amplify and frequency-convert signals so that they reach a suitable state to be converted into digital values, after which further signal processing can be performed in the digital domain.

■ Voltage and Current 电压和电流

The electricity provided by the source has two basic characteristics, called voltage and current.

Voltage, known as electrical potential difference or electric tension , measured in volts, is the difference in electric potential between two points or the difference in electric potential energy per unit charge between two points. The electrical "pressure" causes free electrons to travel through an electrical circuit, also known as electromotive force (EMF).

A voltmeter can be used to measure the voltage (or potential difference) between two points in a system; usually a common reference potential such as the ground of the system is used as one of the points. Voltage can be caused by static electric fields, by electric current through a magnetic field, by time-varying magnetic fields, or a combination of all three.

Electric current is a flow of electric charge through a medium. This charge is typically carried by moving electrons in a conductor such as wire. It can also be carried by ions in an electrolyte or by ions in plasma.

The SI unit for measuring the rate of flow of electric charge is the ampere, which is charge flowing through some surface at the rate of one coulomb per second. That is, electric current is measured using an ammeter.

▪ Ohm's Law 欧姆定律

Many "laws" apply to electrical circuits, but Ohm's Law is probably the most well known. Ohm's Law states that an electrical circuit's current is directly proportional to its voltage, and inversely proportional to its resistance. So, if voltage increases, for example, the current will also increase; and if resistance increases, current decreases.

To understand Ohm's Law, it's important to understand the concepts of current, voltage, and resistance: current is the flow of an electric charge, voltage is the force that drives the current in one direction, and resistance is the opposition of an object to having current pass through it.

The formula for Ohm's Law is E = I x R, where E = voltage in volts, I = current in amperes, and R = resistance in ohms. This formula can be used to analyze the voltage, current, and resistance of electricity circuits.

Notes and Expressions

1. charge [tʃɑ:dʒ] *n.* 费用；电荷；掌管 *vi.* 充电；控告；索价
2. electron [i'lektrɔn] *n.* 电子
3. electrolyte [i'lektrəulait] *n.* 电解液，电解质；电解
4. plasma ['plæzmə] *n.* [等离子] 等离子体；血浆；电浆
5. SI (System International) 国际单位制
6. coulomb ['ku:lɔm] *n.* 库仑（电量单位）
7. load [ləud] *n.* 负载，负荷；工作量 *vt.* 使担负；装填
8. discrete [dis'kri:t] *adj.* 离散的，不连续的 *n.* 分立元件；独立部件
9. hybrid ['haibrid] *n.* 杂种，混血儿；混合物 *adj.* 混合的；杂种的

Section 2 Electronic Component 电器元件

Read the following passages and try to describe these components in English.

An electronic component or electronic element and may be available in a discrete form having two or more electrical terminals (or leads) (as Figure 9-2 shows). These are intended to be connected together, usually by soldering to a printed circuit board, in order to create an electronic circuit with a particular function , for example an amplifier, radio receiver, or oscillator. Electronic components may be packaged discretely or integrated inside of packages like film devices.

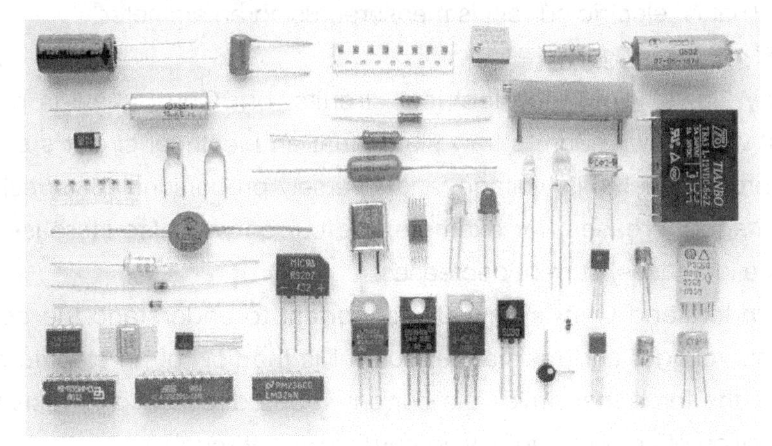

Figure 9-2 Electronic Components (电子元件)

▪ Diodes 二极管

Most diodes (as Figure 9-3 shows) are similar in appearance to a resistor and will have a painted line on one end showing the direction or flow (white side is negative).

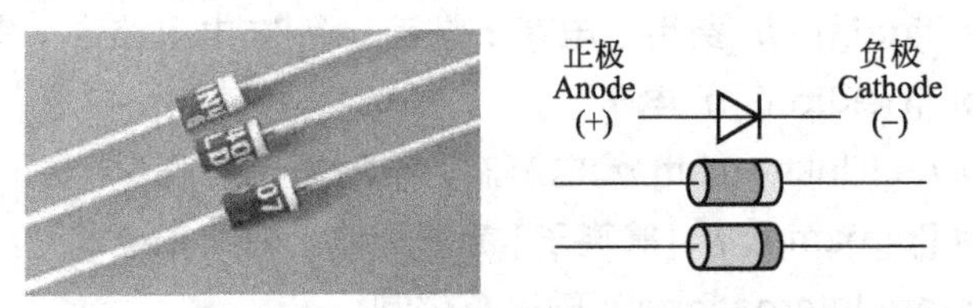

Figure 9-3 Diode and Symbol (二极管及其符号)

Diodes are basically a one-way valve for electrical current. They let it flow in one direction (from positive to negative) and not in the other direction. If the negative side is on the negative end of the circuit, current will flow. If the negative is on the positive side of the circuit, no current will flow.

▪ Light-emitting Diode 发光二极管

A light-emitting diode (LED) (as Figure 9-4 shows) is an electronic light source. LEDs are based on the semiconductor diode. When the diode is forward biased (switched on), electrons are able to recombine with holes and energy is released in the form of light. This effect is called electroluminescence and the color of the light is determined by the energy gap of the semiconductor.

LED presents many advantages over traditional light sources including lower energy consumption, longer lifetime, improved robustness and smaller size.

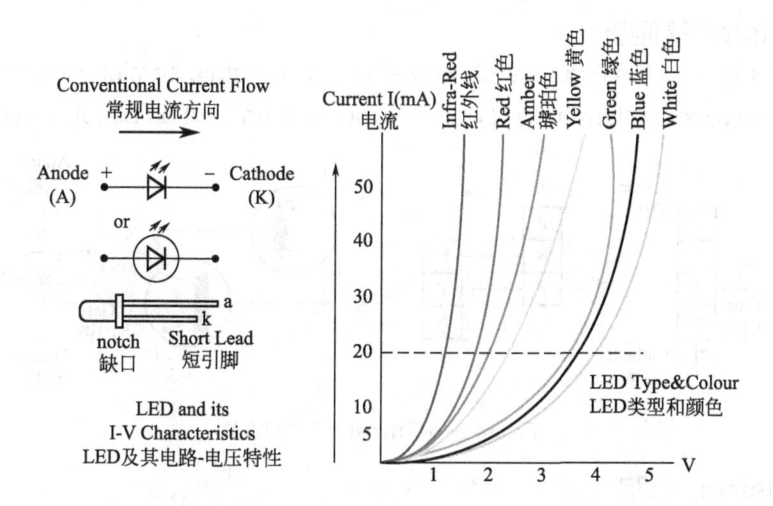

Figure 9-4 LED（发光二极管）

▪ Transistors 晶体管/三极管

The transistor (as Figure 9-5 shows) is the key component in all modern electronics. Many consider it to be one of the greatest inventions of the 20th century. There are two types, NPN and PNP, with different circuit symbols.

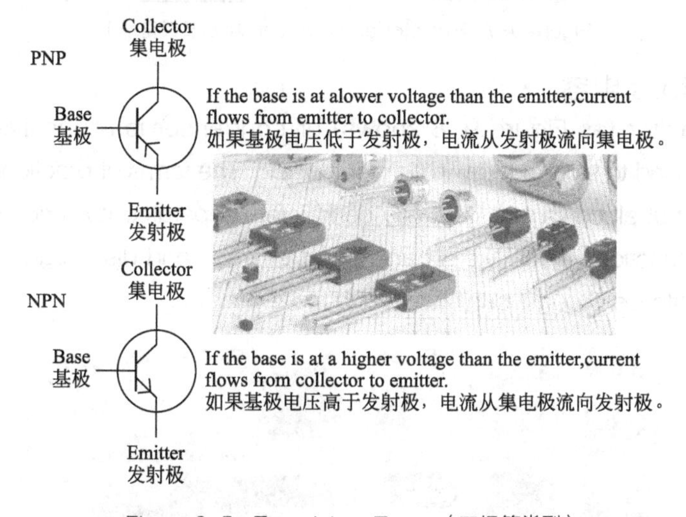

Figure 9-5 Transistors Types（三极管类型）

The essential usefulness of a transistor comes from its ability to use a small signal applied between one pair of its terminals to control a much larger signal at another pair of terminals. This property is called gain. A transistor can control its output in proportion to the input signal. That is, it can act as an amplifier. The transistor can be used to turn current on or off in a circuit as an electrically controlled switch.

- **Thyristor 晶闸管**

The thyristor (as Figure 9-6 shows) is a solid-state semiconductor device with four layers of alternating N and P-type material, act as bistable switches.

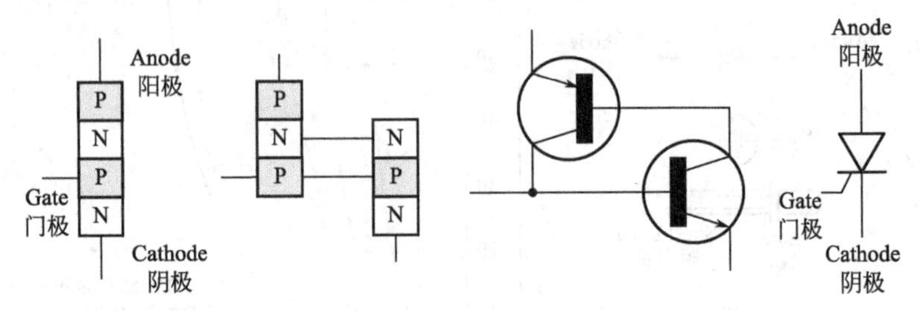

Figure 9-6 Thyristor（晶闸管）

- **Resistors 电阻**

Resistors (as Figure 9-7 shows) "resist" the flow of electrical current. The higher the value of resistance (measured in ohms) the lower the current will be, with the voltage being constant.

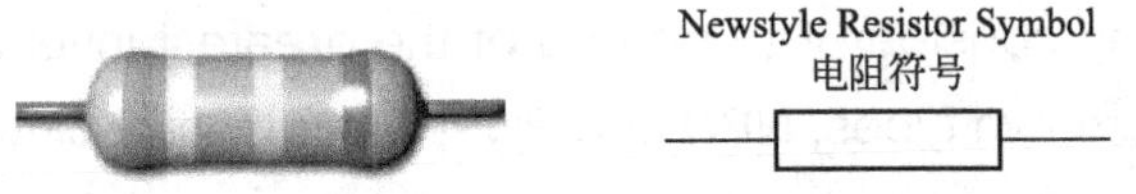

Figure 9-7 Resistor and Symbol（电阻及其符号）

- **Capacitors 电容**

A capacitor (as Figure 9-8 shows) or condenser is a passive electrical component used to store energy in an electric field. The forms of practical capacitors vary widely, but all contain at least two conductors separated by a non-conductor. A capacitor functions much like a battery, but charges and discharges much more efficiently (batteries, though, can store much more charge).

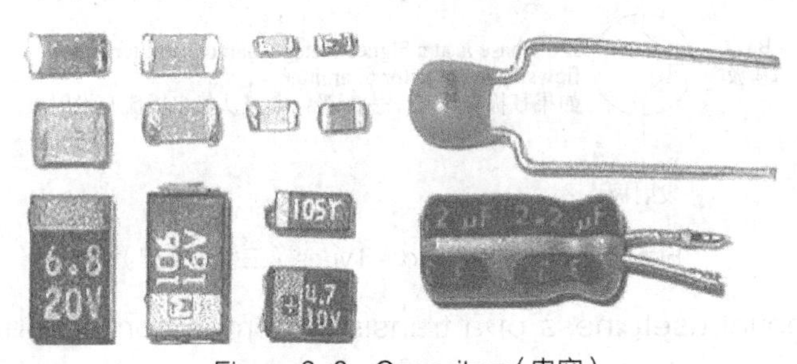

Figure 9-8 Capacitors（电容）

- **Relay 继电器**

A relay is an electrically operated switch. Current flowing through the coil of the relay creates a magnetic field which attracts a lever and changes the switch contacts. The oil current can be on or off so relays have two switch positions and they are double throw (changeover) switches.

The first relays were used in long distance telegraph circuits, repeating the signal coming in from one circuit and re-transmitting it to another. Relays were used extensively in telephone exchanges and early computers to perform logical operations.

▪ MCU/ Single Chip Microprocessor 单片机

Single chip microprocessor (as Figure 9-9 shows) is a complete computer system integrated on a single chip. Even though most of its features in a small chip, it has the majority of computer components: CPU, memory, internal and external bus system. At the same time, it also integrates communication interfaces, timers, real-time clock and other peripheral equipment. And now the most powerful single chip microcomputer system can even integrate voice, imagine, networking, input and output complex system on a single chip.

Figure 9-9 Single Chip Microprocessor(单片机)

If you want to get some English information about single chip microprocessor in the internet, what do you think should be the key words, single chip microprocessor, SCP, microcontroller unit, Microcontroller or MCU? Since "Microcontroller unit" is more popular than "single chip microprocessor" and abbreviation usually have more than one meaning, the best key words are "Microcontroller unit" and "Microcontroller" . As you know, suitable key word can result in narrow range and benefit you with information you want.

▪ PLC 可编程控制器

A programmable logic controller (PLC) or programmable controller is a digital computer used for automation of electromechanical processes, such as control of machinery on factory assembly line. Unlike general-purpose computers, the PLC is designed for multiple inputs and output arrangements, extended temperature ranges, immunity to electrical noise, and resistance to vibration and impact.

The PLC (as Figure 9-10 shows) was invented in response to the needs of the American automotive manufacturing industry. Programmable logic controllers were initially adopted by the automotive industry where software revision replaced the re-wiring of hard-wired control panels when production models changed.

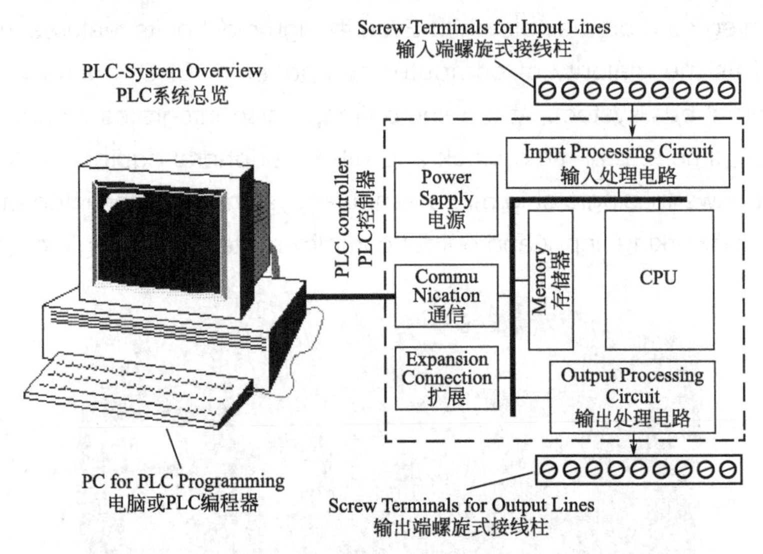

Figure 9-10 PLC（可编程控制器）

Before the PLC, control, sequencing, and safety interlock logic for manufacturing automobiles was accomplished using hundreds or thousands of relays, cam timers and dedicated closed-loop controllers. The process for updating such facilities for the yearly model change-over was very time consuming and expensive, as electricians needed to individually rewire each and every relay.

In 1968 the automatic transmission division of General Motors issued a request for proposal for an electronic replacement for hard-wired relay systems. The first PLC, designated the 084 because it was Bedford Associates' eighty-fourth project, was the winning proposal. Bedford Associates started a new company dedicated to developing, manufacturing, selling, and servicing this new product: Modicon, which stood for "modular digital controller".

One of the people who worked on that project was Dick Morley, who is considered to be the "father" of the PLC. The Modicon brand was sold in 1977 to Gould Electronics, and later acquired by German Company AEG and then by French Schneider Electric, the current owner.

Notes and Expressions

1. discrete [dis'kri:t] *adj.* 离散的，不连续的 *n.* 分立元件；独立部件
2. condenser [kən'densə] *n.* 冷凝器；电容器；聚光器
3. dielectric [ˌdaii'lektrik] *adj.* 非传导性的；诱电性的 *n.* 电介质；绝缘体
4. electroluminescence [i'lektrəuˌlju:mi'nesəns] *n.* 场致发光电，电致发光
5. anode and cathode 阳极和阴极
6. lead [li:d] *n.* 领导；铅；导线；引线；榜样
7. hard-wired ['hɑ:d'waiəd] *adj.* 硬连线的，硬接线的

【Self Evaluation】/【自我评价】

1. Diodes are basically a two-way valve for electrical current.

(A) True (B) False

2. What does this stand for ( )

A
G
C

(A) thyristor (B) PNP (C) NPN (D) capacitor

3. ____________ is a complete computer system integrated on a single chip.

(A) PLC (B) MCU (C) Relay (D) LED

4. The first PLC was designated the 084 because it was very secret.

(A) True (B) False

5. The last owner of Modicon is an(a) __________ company.

(A) American (B) German (C) French (D) China

6. The SI unit for measuring current is ________.

(A) Volt (B) Ohm (C) Ampher (D) Coulomb

7. A battery functions like a capacitor, but charges and discharges much more efficiently.

(A) True (B) False

8. Most diodes are similar in appearance to a resistor.

(A) True (B) False

9. The PLC was invented in response to the needs of _____.

(A) Aircraft (B) Elevator (C) Computer (D) Automobile

10. Relays have ________ switch position(s).

(A) one (B) two (C) three (D) four

【Extensive Reading】【拓展阅读】

Read the following material and retell in your own language.

Conversation About Circuit Elements 关于电路元件对话

In this conversation, E is the shortening of Engineer; S is the shortening of Students.

E: Let's begin by talking about three types of circuit elements: capacitors, resistors and inductors.

S: Are resistors of those circuit elements which have very high resistance values?

E: Not necessarily. A resistor is a kind of circuit element whose resistance is known within certain tolerance limits. They are used in circuits to realize required resistance values.

S: But they do have capacitance and inductance values. You said that all substances have these properties.

E: Yes, they do have low levels of stray capacitance and inductance, but these are usually negligible in practical applications.

S: what are capacitors and inductors?

E: A capacitor is a circuit element made up of two metal plates separated by a non-conducting material. An inductor is a circuit element make a coil of wire. However, for circuit applications their current-voltage relationships are more important.

S: Oh. Do capacitors and inductors behave like resistors?

E: In a way. Both capacitors and inductors display a resistance-like quality called reactance. However, reactance is frequency-dependent. For DC a capacitor has infinite reactance while an inductor has zero reactance.

S: Would you please define reactance?

E: It is the ratio of AC voltage to AC current.

S: This sounds exactly like resistance to me.

E: Yes, but it is frequency-dependent. Resistance is frequency-independent. At very high frequencies a capacitor has extremely low reactance, and an inductor has extremely high reactance.

S: So they display opposite current-voltage relationships with respect to frequency variations. In DC a capacitor will have a voltage across it but no current through it, while an inductor has no voltage across it but a current flowing through it.

E: That's right.

S: What are the specific current-voltage relationships?

E: The capacitance of a circuit element is defined as the ratio of current flowing through it to the time derivative of the voltage across it. Inductance is defined as the ratio of the voltage across the inductor to the time derivative of the current flowing through it.

S: Time derivatives. Now I see why reactance changes as frequency changes. Do these elements have any DC applications?

E: Very large capacitors store very large charges and are used in a number of different applications. Very large inductors are often used as extremely powerful electromagnets.

S: Both capacitors and inductors store electrical energy. Are there any circuit elements which supply other types of energy?

E: Of course. We have devices such as batteries which convert chemical energy into DC voltage. Commercial power generators supply AC voltages by converting mechanical and nuclear energy into electrical energy.

S: Resistors don't store or supply energy; so I guess they must dissipate it.

E: That's right. The energy dissipated by a resistor gets spent by being heated.

Notes and Expressions

1. inductance [in'dʌktəns] *n.* 电感；感应系数；自感应
2. reactance [ri'æktəns; ri:-] *n.* 电抗
3. infinite ['infinət] *adj.* 无限的；无穷的；极大的；无数的
4. differential [ˌdifə'renʃəl] *adj.* 微分的；差别的 *n.* 微分；差别
5. derivative [di'rivətiv] *n.* 衍生物；派生物；导数 *adj.* 派生的；引出的
6. dissipate ['disipeit] *vt.* 浪费；使……消散 *vi.* 驱散；放荡

Chapter 10

Installation and Maintenance
电梯安装与维保

【Content Description】/【内容描述】

电梯安装前应该熟知安装相关的法规和标准，安装过程是电梯知识的综合运用，通过安装可以加深对零部件的认识，了解其安装位置和使用要求。

维保是一个复杂的过程，作为一个合格的电梯技术员必须了解这些过程，并在每个环节保证安全。如果未能按照上述要求操作，可能威胁到工人本身或公众的安全，当事人根据相关条例会受到政府罚款或承担相关责任。

本章主要讲述了电梯行业安装和施工的关键步骤及安全事项，共三节，包括电梯安装的准备和确认工作，电梯安装的安全注意事项和维保的基本知识。

首先总体讲述了在工地进行施工的总体原则，接下来针对具体对象详细描述了电梯的使用，仿型轿厢、工作平台、钢丝绳、手拉葫芦的使用方法及注意事项，氧气瓶、乙炔瓶在切割和焊接中的保护措施，绳头固定方式等。最后对维保做了简要介绍。通过学习可以掌握常见的安装词汇，安装流程和注意事项的表达方式。

【Related Knowledge】/【知识准备】

电梯安装是一个复杂的过程，作为一个合格的电梯技术员必须了解这些过程，并在每个环节保证安全。电梯的安装规程包括基本规程，安全用电规程，井道作业规程，吊装作业规程和防火措施规程等。在安装过程中应树立安全第一的意识，不遵守相关的规定会造成安全隐患，造成财产损失，不仅当事人和公司依法会受到处罚，还可能危及员工和他人的人身安全。

国内标准，电梯可以参考《电梯制造与安装安全规范》GB 7588—2003，扶梯可以参考《自动扶梯和自动人行道的制造与安装安全规范》GB16899—2011。国外标准，扶梯可以参考欧洲标准EN115-1:2008，电梯可以参考欧洲标准EN 81-1:1998/A1:2005: E或美国标准A17.1—2007《电梯和自动扶梯安全规范》。

维保的基础是GB7588—2003《电梯制造与安装安全规范》，电梯维修也是使电梯通过维修达到制造安装时的相应标准。电梯维修工可以采用询问、手摸、耳听、目视等方法判断电梯状态。

维保是电梯行业的一个工种，包括日常的维护和一些规模不等的维修。维护亦称为保养，是指在电梯交付使用后，为保证电梯正常及安全运行，而按计划进行所有必要的操作，如润滑、检查、清洁、调整及更换易损件。维修主要包括常见故障的排除和电梯系统结构维修。维修中的分析判断错误，会造成维修时间延长，甚至给电梯带来损失，危及乘客安全。所以，维修人员必须持证上岗。

【Section Implement】/【章节内容】

Section ❶ Elevator Installation 电梯安装

Read the following passages and try to describe the process of elevator installation.

When installing a new elevator, installers begin by studying blueprints to determine the equipment needed. After preparation and verification, they will start to install.

Elevator installers put conduits along a shaft's walls. Once the conduit is in place, mechanics pull plastic-covered electrical wires through it. They then install electrical components and related devices required at each floor and in the machine room.

Installers bolt or weld steel rails to rail bracket on the walls of the shaft to guide the elevator. Elevator car is usually installed at the top of the shaft, and guide shoes and safety gear are attached to the car frame. Installers also install the outer doors and door frames at the elevator entrances on each floor.

The most highly skilled elevator installers, called "adjusters," specialize in fine-tuning all the equipment after installation. Adjusters make sure that an elevator works according to specifications and stops correctly at each floor within a specified time. Adjusters need a thorough knowledge of electronics, electricity, and computers to ensure that newly installed elevators operate properly.

▪ General Verification 总体确认

Verify the following before commencing the installation:

① Check the hoistway is square and plumbed;

② Check the actual dimensions of the hoistway against the general layout drawing supplied, such as width and depth;

③ Check the total travel (distance from floor level to floor level);

④ Check the pit depth;

⑤ Check overhead space (the measurement from the top floor level to the underside of the ceiling);

⑥ Check the fixing points in hoistway where rail brackets mount.

▪ Delivery of Material 物料交付

Compare all components in the shipment against the shipping list to check if there is any shortage. Any damage to crating should be reported to carrier and vendor immediately and a memo sent to them as soon as possible listing the damaged equipment.

■ Installation Schedule 安装进度

The time taken to install these units will vary from project to project and will be dependent upon the following factors.

① The number of floors served;

② The orientation of the unit;

③ If auto or manual door operation is selected;

④ The site conditions and the experience of the installing team;

⑤ Whether entrances are installed by elevator contractor or by others.

It is very important that supervisors visit the site to verify all dimensions are correct and all services are available before the installing crew commences. If scaffold and power are not available then the job should not be started. Normally scaffold is required on day one and power is required to run the machine at the end of day one or beginning of day two.

Notes and Expressions

1. commence [kə'mens] *vi.* 开始 *vt.* 使开始；着手
2. verification [ˌverifi'keiʃən] *n.* 确认，查证；核实
3. scaffold ['skæfəuld] *n.* 脚手架；鹰架；绞刑台

Section ❷ Safety During Installation 装梯安全

Read the following passages and try to list the risk during installation.

■ General Safety Caution 总体安全注意事项

The technician-in-charge has specific responsibility for coordinating jobsite safety at each location, although, safety is everyone's responsibility. Keep work areas clean of waste materials, rubbish and other debris that may cause a tripping hazard. Wear leather gloves and use broom and dustpan when cleaning up to avoid lacerations and punctures from sharp objects that may be hidden in the trash.

Place oily rags in an approved receptacle and have removed from the jobsite periodically to reduce fire hazards. Remove or bend nails in boards and be aware of tripping hazards. Stack used material out of the way so that it can be identified for removal. Damaged rope shall not be used and removed from the jobsite.

At the jobsite, safety is of paramount importance, therefore, the personal use of radios and tape players is not allowed. Naturally, horseplay, practical jokes, fighting and possession of weapons are prohibited.

Employees shall not work when ability or alertness is impaired by:

① Fatigue;

② Intoxicating beverages;

③ Illegal or prescription drugs;

④ Other causes that would reduce their attention or inhibit their ability.

Where construction or modernization work is done on an elevator in a multiple hoistway with an adjacent elevator operating, the portion of the hoistway where the work is being done shall be separated.

Where the general public is present, solid barricades at least eight feet high or to the ceiling shall fully enclose the work area. Report damaged or improperly maintained guardrails and barricades to your supervisor and general contractor superintendent immediately. Refer to the safety handbook for additional information on guardrail and barricade requirements.

Working safely is the most important part of the job. Safety is everyone's responsibility. Do not hesitate to caution anyone performing an unsafe act. Likewise accept such caution from others with a spirit of thanks. Continuous review of your Elevator Industry Field Employees' Safety Handbook, company safety policy and procedures and specific jobsite requirements are an essential part of your safety responsibility.

- Handle with Circuit 电路操作

Always disconnect the circuit, and then use a meter to check for voltage, before doing any work that requires contact with conductors, such as replacement of electrical components. Also be aware that on a multiple-car installation, there may be voltage on a controller even when the mainline disconnect for one car is open.

Never work on any circuit while standing in water or on grounded metal surface. Remember, the elevator mainline disconnect does not remove power from the pit light, sump pump or car top and car lighting.

Never carry tools in pockets. Doing so increases the risk of tools contacting live electrical circuits, snagging on moving equipment or falling on workers below.

When working near or on live electrical circuits, you should:

① Stand on dry board or rubber mat;

② Never stand on metal or wet surfaces;

③ Use only insulated tools and work lights;

④ Remove all jewelry and all key rings;

⑤ Wear safety glasses with non-metallic frames;

⑥ Use fuse puller and verify that power is disconnected when installing fuses;

⑦ Never use jumpers across fuses;

⑧ Replace covers when work is complete.

⑨ Place a sticker in the controller near the inspection switch that reads as follows: "GATE OR SAFETY CIRCUITS SHALL NEVER BE JUMPED OUT."

▪ Labor Insurance Protection 劳保防护

The separation shall be constructed as indicated in the drawing. Company-issued hard hats shall be worn with the suspension properly fitted and located on the head.

Appropriate eye protection shall be worn when:

① Welding;

② Cutting;

③ Babbitting;

④ Using chemicals or solvents;

⑤ Working in dusty areas;

Wear gloves when there is a potential hazard such as handling heavy or rough material. Always wear gloves when handling wire rope. Never wear gloves near moving or rotating machinery or while placing rollers under a load.

▪ Ladders 梯子

Only company-issued ladders with safety feet shall be used. Metal ladders conduct electricity and shall not be used. Ladders shall never be painted because paint may cover up cracks or other defects. Damaged or cracked ladders shall be tagged "DO NOT USE" and removed from the jobsite.

Stepladders shall be fully open and used properly. When a ladder is used where it may be struck by others in the area, there shall be a second person at the bottom of the ladder at all times. Never leave the ladder unattended in such a location. When ladders are placed in aisles or corridors, they shall be barricaded or roped off.

When working from ladder, avoid reaching more than an arm's length. Leaning out to reach can cause imbalance and puts you in a precarious position, risking a fall. Never step on the top step of a ladder.

▪ False Cars 仿形轿厢

There are a variety of false cars and similar devices used in the hoistway to stack rails, set brackets, install entrances, run wiring and other work necessary to install an elevator. These are company tools and their use and testing is dictated by company policy.

Common rules for use of false cars:

① Conduct static safety test daily before use, and check overhead support, all attachment and rigging daily.

② Document all safety tests including those done when the car is placed in service or moved to a different hoistway.

③ Use only to carry elevator workers, and do not use to hoist heavy objects.

④ Be aware of the limited capacity and never exceed it.

⑤ Persons working on a false car shall wear full body harness and shock-absorbing lanyard secured to an independent lifeline.

⑥ Where overhead protection is required, debris net with a one-fourth inch opening is recommended.

▪ Fall Protection 坠落防护

In some instances, while working on ladders, false cars and in other areas of construction and service, a personal fall arrest system is required. Fall protection must be used when a worker is exposed to a fall hazard while working more than 6 ft (1.8 m) above a lower level and an opening of more than 10 in. (254 mm).

Guardrail systems must be used when working in and around the hoistway. This is for your protection as well as the protection of others that may be working within your area. Equipment that has been subjected to fall impact shall be taken out of service and destroyed.

The following are a list of rules which, must be followed when using a fall arrest system:

① Only company approved lifelines, shock absorbing lanyards and body harnesses shall be used;

② Lifelines shall be protected against being cut or abraded;

③ Only one worker shall be permitted on a vertical lifeline;

④ Tying directly to the hoist line is prohibited, a proper rope grab must be used;

⑤ Lifelines shall be installed before working in the hoistway, shall be run the full length of the hoistway.

- **Chain Hoist 手拉葫芦**

Visually inspect chain hoist daily for defects prior to use. Look closely at the chain and lower hook. The lower hook is the weakest part of the hoist. Damage or spreading of the lower hook is an indication of overloading. Measure the spread of the lower hook and compare with the maximum allowable as shown in the safety handbook. Never replace, lengthen or splice the chain or exceed the capacity of the hoist. Repairs shall be made only by the manufacturer or authorized agent.

- **Gas Tanks 气罐**

Acetylene and oxygen tanks are under high pressure and shall be handled with extreme care. Store oxygen and acetylene tanks in the upright position, at least 20 feet apart or separated by a one-half hour fire resistant wall at least five feet high when not in actual use.

Never lubricate the threads on gauges or tanks. Take care with the hose to prevent it from being damaged and do not use frayed hoses.

- **Precautions for Welding 焊接防护措施**

Before starting to weld, clean the area of all combustible materials. If combustible materials cannot be moved, cover them with fire-retarding material. Be especially aware of oil or solvents. Wood floors shall be wet down or covered with sheet metal or equivalent material. Where sparks are likely to fall should be noted since they may start a fire or fall on other workers.

Wear proper clothing for welding. Never wear oil-stained clothing or clothing that ignites and burns easily. An additional person shall stand by with an "ABC" fire extinguisher to watch for fire where welding is in process. The area and adjacent areas shall be monitored for a while after the welding is complete.

Proper eye protection for the welding operation is very important. Welder's helmet with head protection shall be used for electric arc welding.

- **Rope Fastening 钢丝绳紧固**

It is recommended that wedge shackles be used where possible. However, if tapered sockets are used, observe the following precautions:

① Wear face shields and gloves when pouring babbitt.

② Preheat the socket to be sure that there is no moisture present before pouring the melted babbitt into the socket.

③ Avoid breathing fumes and work only in a well-ventilated area.

④ Wash hands thoroughly prior to eating or smoking after handling babbitt.

When attaching wire rope with rope clips, refer to the safety handbook for the required number of clips and the distance between clips. Place the first clip one clip saddle width from the dead end of the rope. Apply the "U-bolt" on the dead side of the rope, the live side resting in the saddle.

Tighten the nuts evenly, alternating from one nut to the other until the recommended torque is reached. The recommended torque can be found in the wire rope users manual or clip supplier catalog. It is usually 45 ft-lbf for a three-eighth-inch rope and 65 ft-lbf for one-half inch rope. The threads shall be clean, dry and free of lubricant for accurate torque. Place the expected load on the line and check the connections.

Over-tightening will distort and damage the rope, weakening the connection. Likewise, under-tightening may allow the rope to slip, damaging it and weakening the connection. After the rope connection has been in service a few days, inspect it for signs of movement or damage and tighten to the recommended torque value.

Notes and Expressions

1. debris ['deibri:] *n.* 残骸；碎片；瓦砾，废墟；垃圾；破片；残渣；废弃物

2. hazard ['hæzəd] *n.* 危险；冒险；危险的根源；引起(或导致)危险的事物

3. laceration [ˌlæsə'reiʃən] *n.* 撕裂；扯破；割碎；裂口；伤口

4. broom [bru:m] and dustpan ['dʌstpæn] 扫帚和簸箕

5. receptacle [ri'septəkl] *n.* 容器；插座

6. paramount *adj.* 最重要的，主要的；至高无上的 *n.* 最高统治者

7. horseplay ['hɔ:splei] *n.* 嬉戏；胡闹；喧闹；恶作剧

8. babbitt ['bæbit] *n.* 巴氏合金 *vt.* 给……浇巴氏合金

9. combustible [kəm'bʌstəbl] *adj.* 可燃的；易燃的，暴躁的 *n.* 可燃物；燃料

10. shackle ['ʃækl] *n.* 束缚；桎梏；脚镣 *vt.* 束缚；加枷锁

11. prescription [pris'kripʃən] *n.* 药方；指示；惯例 *adj.* 凭处方方可购买的

12. precarious [pri'kεəriəs] *adj.* 不稳定的；靠不住的；不安全的，危险的

13. abrade1 [ə'breid] *vt.* 擦掉；摩擦，磨蚀，磨损

Section ❸ Elevator Maintenance 电梯维保

Read the following passage and give a short description in your own words.

Once an elevator is put into operating, it must be maintained and serviced regularly to keep it in safe working condition. Elevator repairers generally do preventive maintenance—such as oiling and greasing moving parts, replacing worn parts, testing equipment with meters and gauges, and adjusting equipment for optimal performance. They ensure that the equipment and rooms are clean. They also troubleshoot and may be called to do emergency repairs. People who specialize in elevator maintenance work independently most of the day and typically service many of the same elevators on multiple occasions over time.

Service crews also need to handle major repairs—for example, replacing cables, elevator doors, or machine bearings. These tasks may require the use of cutting torches or rigging equipment—tools that an elevator repairer would not normally carry. Service crews also do alteration work, such as moving and replacing electrical motors, hydraulic pumps, and control panels.

【Self Evaluation】/【自我评价】

1. Ladders used in elevator installation shall ________ be painted.

(A) usually (B) never (C) occasionally

2. In chain hoists, damage or spreading of the lower hook is evidence of__________.

(A) weak metal (B) heat strain (C) overloading

3. When working in a location more than __________ feet, a full-body harness is required.

(A) 10 (B) 2

(C) 7 (D) 6

4. Safety is only supervisor's responsibility. (　　)

(A) True (B) Fasle

5. We can not replace or lengthen the hoist chain by ourselves. (　　)

(A) True (B) Fasle

6. ________ part is the weakest part of the hoist.

(A) Top (B) Lower (C) Middle

7. Always wear _________ when handling wire rope.

(A) gloves (B) glasses (C) eye shield

8. Compare components in shipment against ________.

(A) shipping list (B) the contractor

(C) carrier (D) vendor

9. Glasses with __________ can be worn when working on live circuit.

(A) shipping list (B) the contractor

(C) carrier (D) vendor

10. The recommended torque is usually ______ ft-lbf for a three-eighth-inch rope.

(A) 30 (B) 45

(C) 35 (D) 65

11. Apply the "U-bolt" on the __________side of the rope, the __________ side resting in the saddle.

(A) live (B) big (C) small (D) dead

12. During welding, wood floors shall be wet down or covered with sheet metal or equivalent material. (   )

(A) True (B) Fasle

13. When working from ladder, avoid reaching more than an arm's length. (   )

(A) True (B) Fasle

14. Place oily rags in an approved receptacle and have removed from the jobsite periodically to reduce fire hazards. (   )

(A) True (B) Fasle

15. The most highly skilled elevator installers, called _________.

(A) service screw (B) repairer (C) adjuster (D) installer

【Extensive Reading】/【拓展阅读】

Special Elevator
特殊电梯

▪ TWIN: 2 Cabs, 1 Shaft, 0 Crowds 孪生电梯

Two cars running in one shaft is an idea realized by a combination of expertise and state-of-the-art technology. The first TWIN elevator system was installed at Stuttgart University in 2002.

Two cars arranged one above the other can run independently and also at different speeds? In the same shaft to within a set safety distance, this depends on the speeds involved. The cars can move in different directions which means that they can also move toward each other. Each car has its own traction sheave drive and counterweight, but both use the same guide rails.

A four-level TÜV-certified safety system ensures optimum comfort, reliability and safety at all times. The TWIN system is designed for buildings with a minimum height of 50 meters. It is used in combination with at least one conventional elevator to make it possible to travel from the ground to the top floor.

The optimum configuration is with two or more main access stops. This elevator system opens up completely new concepts for mobility in buildings.

▪ Elevator Works without Cables 无线电梯

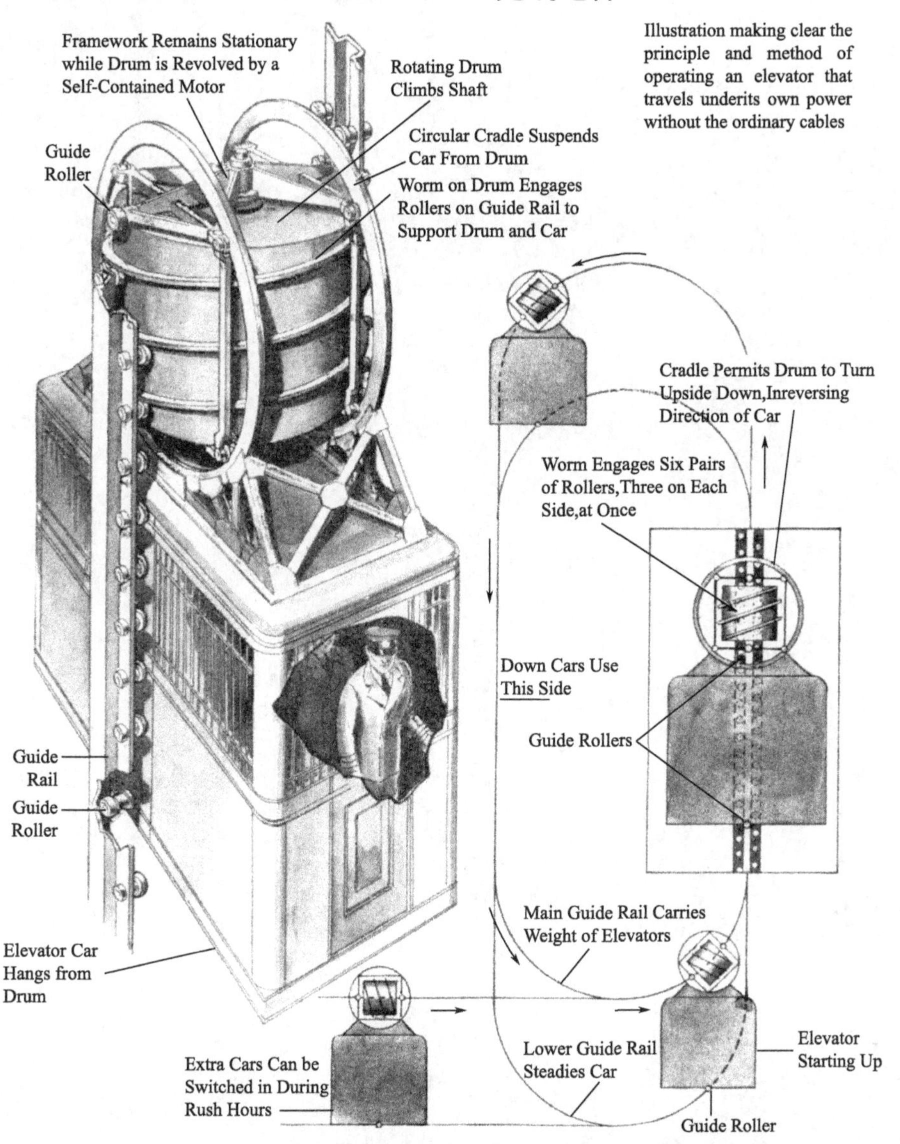

New Elevator Works without Cables

Elevators travel under their own, self-contained power in a system upon which a New York inventor has just received a patent. Each car is suspended from a hollow drum containing a driving motor. Under control of the operator, the drum revolves and climbs a vertical series of rollers by means of a worm on its exterior, as shown in the diagrams. Reaching the top of its endless shaft, the drum inverts itself and starts down the other side, the elevator car remaining upright meanwhile. Advantages of the new system, the inventor declares, are that extra cars may be used during rush hours and withdrawn when not needed; also, that the system removes present restrictions that limit the height of elevator shafts.

Chapter 11

Escalator and Moving Walks
扶梯和自动人行道

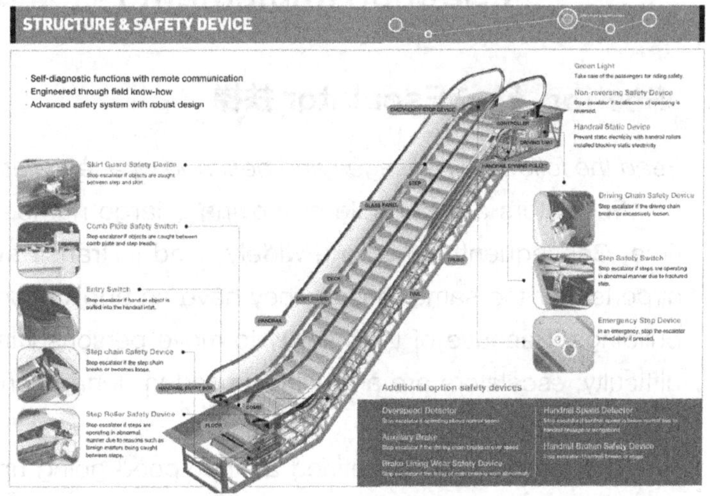

【Content Description】/【内容描述】

本章对扶梯和自动人行道进行了介绍，配有大量的插图和表格进行说明，内容包括扶梯和自动人行道的结构，相关的零部件，安全装置和相关标准。本章内容对成为一名合格的扶梯技术人员至关重要，因为这些知识是扶梯技术人员日后工作将会遇到的问题。

本章知识有助于对扶梯和自动人行道进行整体的理解，而相关安全装置的知识对于扶梯与自动人行道技术人员也非常重要，以确保系统可以根据要求进行维保，从而确保公众人身和财产的安全。

【Related Knowledge】/【知识准备】

扶梯一般是斜置，常用角度为30°和35°。行人在扶梯的一端站上自动行走的梯级，便会自动被带到扶梯的另一端，途中梯级会一路保持水平。扶梯在两旁设有跟梯级同步移动的扶手，供使用者扶握。电动扶梯可以是永远向一个方向行走，但多数都可以根据时间、人流等需要，由管理人员控制行走方向。

自动扶梯的核心部件是两根链条，它们绕着两对链轮进行循环转动。在扶梯顶部，有一台电动机驱动传动齿轮，以转动链圈。发动机和链条系统都安装在桁架中，桁架是指在两个楼层间延伸的金属结构。

另一种和电动扶梯十分类似的行人运输工具是自动人行道，角度一般为0°、6°和12°。两者的分别主要是自动行人道是没有梯级的，多数只会在平地上行走或是稍微倾斜。扶梯一般有四条导轨，而人行道一般是两条导轨。

【Section Implement】/【章节内容】

Section 1 Escalator 扶梯

Read the following passages and describe what escalator is.

Escalators are capable of moving a large number of people in a short time. Consequently, they are widely used in transit systems, subways and airports. For the same reason, they have an application in schools and office buildings. Because of their ability to move persons with ease and little to no difficulty, escalators are also widely used in department stores and shopping malls.

The rated speed is defined as the speed going up, measured along the angle of the incline, with rated load. The most common speed of escalators in the U.S. is between 90–100 ft/min (0.45–0.51m/s). A17.1–2000 (6.1.4.1) reduced the maximum rated speed of escalators from 125 ft/min to 100 ft/min. Escalators in shopping malls and department stores may operate a bit slower than those in transit facilities. Escalators must be started with a key-operated switch, automatic starting by any means is prohibited.

The theoretical capacity of an escalator 40 in. wide traveling at 90 ft/min with one person on every step is 8,102 people per hour while the nominal capacity is approximately half that (Note: Step run is 15.75 inch). They are not intended to move persons with handicaps, baby strollers, luggage carriers or large objects, and even persons with disabilities that impair their ability to walk, such as arthritis, have difficulty boarding and exiting an escalator. Consequently, there is also a need for elevators where escalators are used.

▪ Escalator Parts and Structure 扶梯零件及其结构

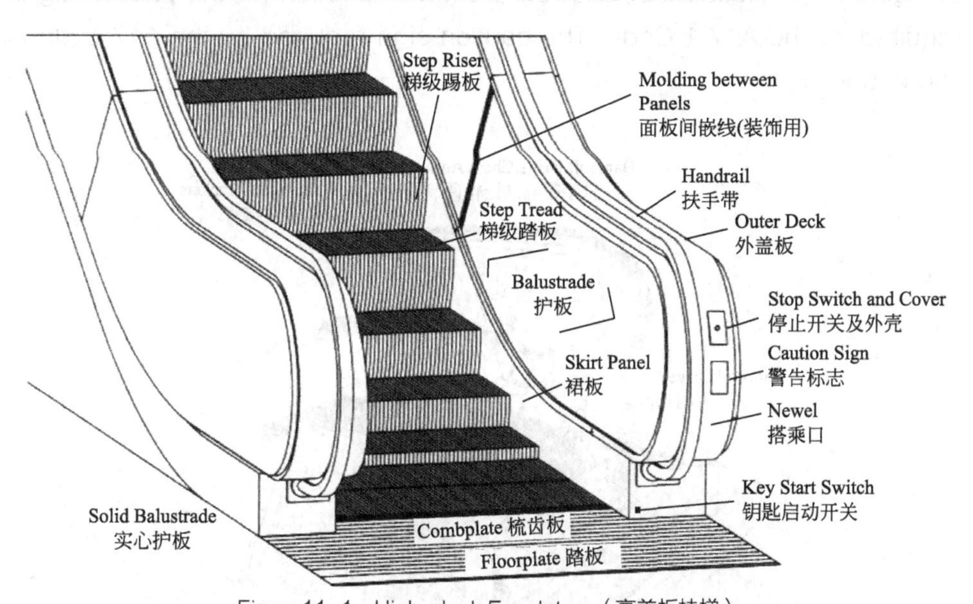

Figure11–1 High-deck Escalators（高盖板扶梯）

The parts of high-deck escalators with solid balustrades are shown in Figure11–1. Notice that the stop button is located in plain view at both ends of the unit. The same component designations apply to moving walks except the steps are referred to as treads. In the past, the stop buttons were located on the front plate near the start switch where they were less visible, with the result that many people did not realize they were there.

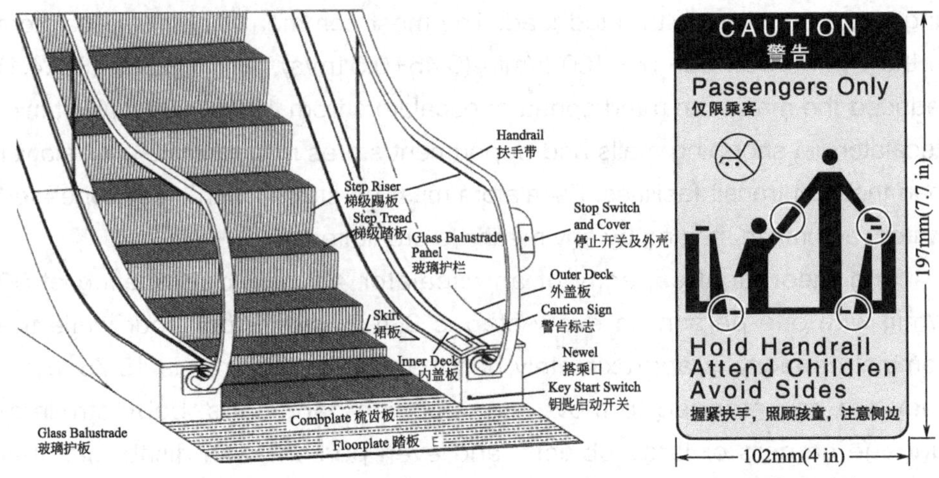

Figure11-2 Low-deck Escalators and Caution Sign（低盖板扶梯与警示标示）

The parts of a low-deck escalator with glass balustrades are shown in Figure11-2. Notice that both units show a location for the caution signs required by the A17.1 Code. The caution sign required by the A17 code is shown above.

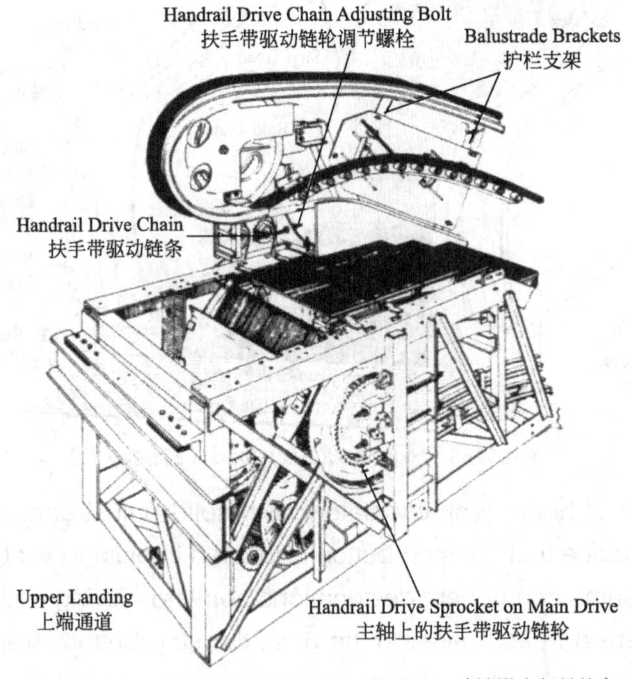

Figure11-3 Internal Parts of an Escalator（扶梯内部结构）

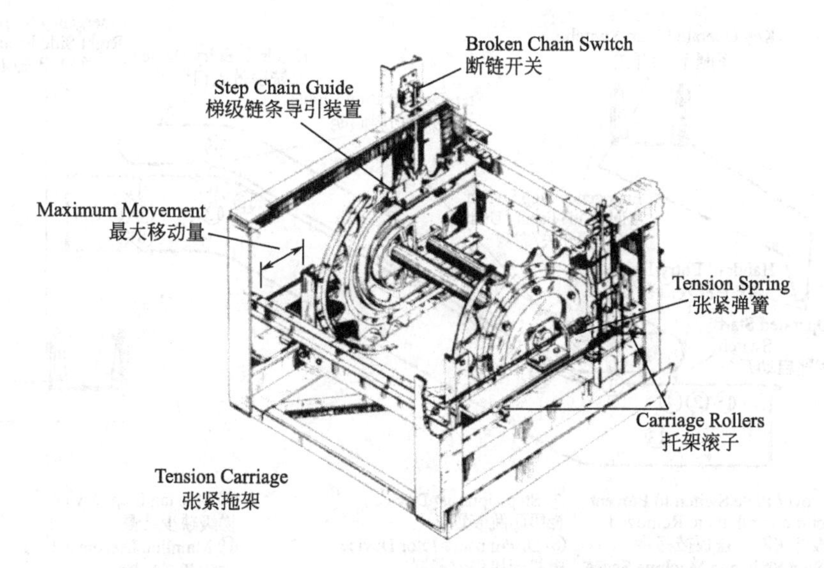

Figure11-4 Internal Parts of an Escalator（扶梯内部结构）

As seen in figure11-3 and figure11-4, an escalator is a very complicated mechanism that requires careful maintenance by trained and qualified personnel.

▪ Safety Devices 安全装置

Escalators are used by thousands of people every day. Responsible building owners all over the world are doing their part to keep people safe by upgrading their existing vertical transportation equipment to meet all local and national safety codes. They provide protection against potential accidents and injuries for passengers with these safety products.

Modem escalators are designed to be as safe as possible, and they have many safety devices that detect unsafe conditions. These devices are shown in figure11-5.

1) The cover plate switch prevents operation of the escalator if the cover plate is removed for maintenance or inspection. That is, the cover plate switch will be triggered if the cover is removed, then the escalator will be power off.

2) The stop switch in the machine space is for the safety of maintenance persons.

3) Skirt obstruction devices are located on both sides on each end. If something is trapped between the steps and skirt as the step approaches the combplate, this switch will stop the escalator before the trapped object contacts the combplate.

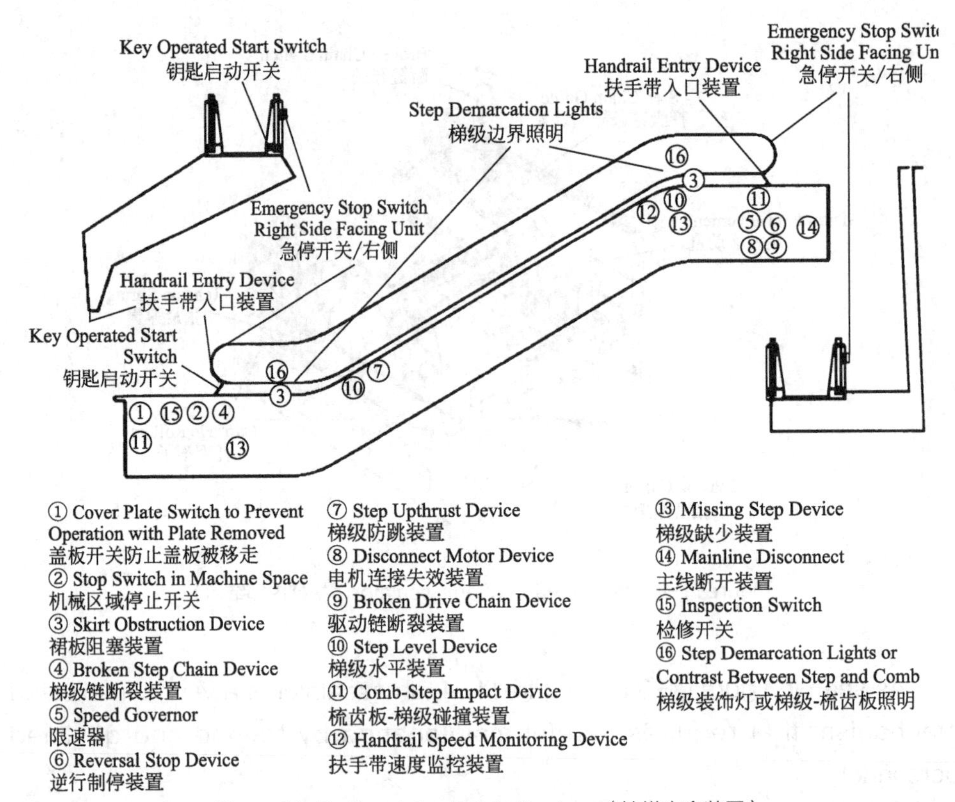

Figure11-5 Escalator Safety Devices（扶梯安全装置）

4) If the step chain breaks, the escalator will stop, this the function of "broken step chain device". In most cases, the stop means the power is off. The step may move with gravity since there is hardly any device like safety gear in elevator.

5) Speed governors will stop the escalator if its speed exceeds a predetermined value. Speed governors are not required where an AC induction motor is directly connected to the driving machine.

6) The reversal stop device will stop the escalator if the direction of motion changed during operation.

7) The step upthrust device will stop the escalator if the step is displaced upward as it approaches the lower curve.

8) If the motor is connected to the reduction gear by other than a continuous shaft, mechanical coupling or toothed gearing, this device will stop the escalator if the coupling means fails. For example, a "D" belt or chain coupling requires this device.

9) If the drive machine is connected to the main drive by a chain, the device will detect a broken chain and stop the escalator.

10) The step level device will detect a downward displacement of 1/8 inch at either side of the riser end. This device will stop the escalator prior to the displaced step entering the combplate.

11) The comb-step impact device will stop the escalator with either a horizontal or vertical force. Horizontal force not greater than 400 lb/ft on either side or 800 lb/ft in the center or vertical force up not greater than 150 lb/ft is permitted.

12) Handrail-speed monitoring device will stop the escalator if either handrail speed varies too much for two seconds. A stopped handrail device was required before the speed-monitoring device.

13) Missing step device will stop the escalator before the space of the missing step emerges from the comb. You can find most safety devices at or near upper landing and lower landing, most are vacant in the middle zone.

14) The mainline disconnect will open all power to the escalator.

15) The inspection switch will allow movement by continuous pressure for maintenance.

16) Step demarcation lights at both landings below the steps or visual contrast between comb and step to alert passenger of pending demarcation.

17) If the escalator is equipped with a rolling shutter to close the escalator, the switch on the shutter will stop the escalator when it begins to close.

18) Handrail Finger Guards: Guards are required where the handrail enters the newel base to prevent fingers and hands from being drawn into the balustrade where the handrail enters. They may be resilient boots, brushes or trap doors. Since 1992, the A17.1 Code has also required a handrail entry device to cause the escalator to stop if an object approaches the area between the handrail and handrail guard or becomes caught in this area.

19) Anti-Slide Devices: Anti-slide devices are required to prevent sliding packages, boxes or persons.

20) Deck Guards: Barricades to prevent access to the outer deck are required to prevent persons from entering and falling.

21) Safety Zone: The A17.1 Code requires a safety zone at entry and exit to be kept clear of obstructions.

22) Escalator safety brushes mount to the skirt panel or inner decking of the escalator are used to alert passengers when they are too close to the gap between the escalator step and skirt panel.

The basic design configuration and dimensions of the escalator are established with safety foremost in mind. Improvements in design and A17.1 Code requirements are constantly being made. One recent change involves means to reduce entrapment between the step side and skirt panel. This establishes a performance index for the clearance and coefficient of friction on the skirt. It also recognizes the use of deflectors to prevent contacting the skirt. The code also recognizes dynamic skirts allowing the skirt to move with the step reducing the possibility of entrapment.

Notes and Expressions

1. pertinent ['pə:tinənt] *adj.* 相关的，相干的；中肯的；切题的
2. incline [in'klain] *vi.* 倾斜；倾向 *vt.* 使倾斜 *n.* 倾斜；斜面；斜坡
3. 1 feet =0.3048m 1ft./min=0.0508m/s 1inch= 25.4mm=0.0254m
4. baby stroller 婴儿推车
5. arthritis [ɑ:'θraitis] *n.* [外科] 关节炎
6. truss [trʌs] *n.* 束；构架；捆；[建]桁架 *vt.* 捆绑；用构架支撑
7. tread [tred] *n.* 踏；鞋底；踏板；梯级 *vt.* 踏；踩；践踏
8. balustrade [ˌbæləs'treid] *n.* 栏杆；扶手；扶栏
9. cleated riser 梯级竖板
10. newel ['nju:əl] 楼梯上下端的栏杆支柱；中心柱
11. induction [in'dʌkʃən] *n.* [电磁] 感应；归纳法；感应现象
12. upthrust ['ʌpθrʌst, ʌp'θrʌst] *n.* 向上推；[地质]地壳隆起
13. demarcation [ˌdi:mɑ:'keiʃən] *n.* 划分；划界；限界
14. resilient [ri'ziliənt, -jənt] *adj.* 弹回的，有弹力的
15. prominent ['prɔminənt] *adj.* 突出的，显著的；杰出的；卓越的
16. soffit ['sɔfit] *n.* 拱腹；下端背面；天花底

Section ❷ Moving Walkway 自动人行道

▪ Moving Walkways Types 自动人行道类型

Moving walks (as shown is Figure 11-6) have many of the same safety design features as do escalators and should be treated in the same manner. Moving walks have a maximum slope of 12 degrees (escalators are usually 30 degrees plus or minus 1 degree). There are two types of moving walks being manufactured today: pallet and continuous belt.

Pallet types escalator consists of many separate pallets connected with a chain, much like an escalator. The pallet may be die-cast aluminum or stainless sheet-metal, widths are between 32 inches (800 mm) and 56 inches (1200 mm). They are usually with a speed of 100 feet per minute (0.5 m/s), powered by an AC induction motor. Continuous belt type: A heavy rubber belt is used and supported on each side with rollers below the belt, referred to as edge support.

Both types of moving walkway have a grooved surface to mesh with combplates at the ends. Also, nearly all moving walkways are built with moving handrails similar to those on escalators.

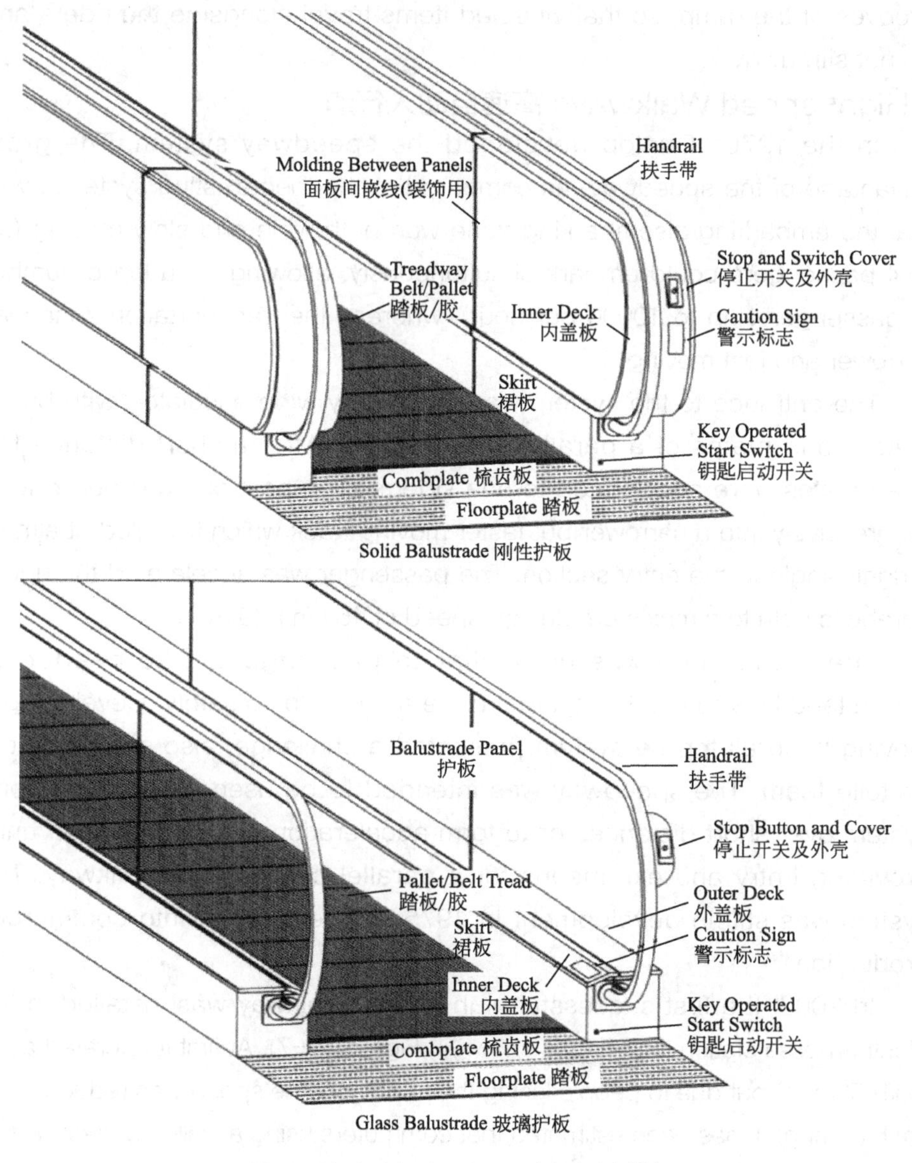

Figure11-6 Moving Walks（自动人行道）

▪ Inclined Moving Walkways 倾斜式自动人行道

An inclined moving walkway is used in airports and supermarkets to move people to another floor with the convenience of an elevator (namely, that people can take along their suitcase trolley or shopping cart, or baby carriage) and the capacity of an escalator.

The carts have either a brake that is automatically applied when the cart handle is released, strong magnets in the wheels to stay adhered to the floor, or specially designed wheels that secure the cart within the grooves of the ramp, so that wheeled items travel alongside the riders and do not slip away.

▪ High-speed Walkways 高速自动人行道

In the 1970s Dunlop developed the speedway system. The great advantage of the speedway, as compared to the then existing systems, was that the embarking/disembarking zone was both wide and slow moving (up to 4 passengers could embark simultaneously, allowing for a large number of passengers, up to 10000 per hour), whereas the transportation zone was narrower and fast moving.

The entrance to the system was like a very wide escalator, with broad metal tread plates of a parallelogram shape. After a short distance the tread plates were accelerated to one side, sliding past one another to form progressively into a narrower but faster moving track which travelled at almost a right-angle to the entry section. The passenger was accelerated through a parabolic path to a maximum design speed of 15 km/h (9 mph).

The experience was unfamiliar to passengers, who needed to understand how to use the system to be able to do so safely. Developing a moving handrail for the system presented a challenge, also solved by the Battelle team. The speedway was intended to be used as a stand alone system over short distances or to form acceleration and deceleration units providing entry and exit means for a parallel conventional walkway. The system was still in development in 1975 but never went into commercial production.

In 2002, the first successful high-speed walkway was installed in the Montparnasse station in Paris (as shown in Figure 11-7). At first it operated at 12 km/h (7 mph) but due to people losing their balance, the speed was reduced to 9 km/h (6 mph). It has been estimated that commuters using a walkway such as this twice a day would save 15 minutes per week and 10 hours a year.

Figure11-7 High-speed Moving Walkway on the Paris Metro
（巴黎地铁的高速自动人行道）

Using the high-speed walkway is like using any other moving walkway, except that for safety there are special procedures to follow when joining or leaving. When this walkway was introduced, staff in yellow jackets determined who could and who could not use it. As riders must have at least one hand free to hold the handrail, those carrying bags, shopping, etc., or who are infirm, must use the ordinary walkway nearby.

Notes and Expressions

1. commuter [kə'mju:tə] *n.* 往返于两地之间的人；定期车票乘客
2. walkway ['wɔ:kwei] *n.* 人行道（尤指有篷的）；通道，走廊
3. infirm [in'fə:m] *adj.*（因年迈而）体弱的；不坚定的；（结构）不牢固的
4. parallel ['pærəlel] *n.* 平行线；对比 *adj.* 平行的；类似的
5. cart [kɑ:t] *n.* 大车；马车；手推车，手拉车
6. slope [sləup] *n.* 斜坡；倾斜 *vi.* 倾斜；逃走

【Self Evaluation】/【自我评价】

1. Escalators are used in transit systems, subways, airports and sports facilities because:

(A) They can move a large number of passengers in a short time.

(B) They can be used for light freight loads for concession stands and luggage needs.

(C) They operate very quietly.

(D) They are easy to maintain.

2. Handrail finger guards:

(A) Will stop operation of the handrail if finger or other objects enter the area.

(B) Operate in both directions.

(C) Operate only if the handrail is exiting the balustrade.

(D) Help prevent fingers and hands from entering the area around the newel base where the handrail enters.

3. Anti-slide devices:

(A) Are required for all escalators.

(B) Are to prevent the steps from sliding from the mounts.

(C) Are to prevent the deck from being used for sliding packages, boxes, persons, etc.

(D) Are never required.

4. Moving walks have a maximum slope of:

(A) 12° (B) 15° (C) 80° (D) 30°

5. The following safety devices are required for moving walks:

(A) Reversal stop device.

(B) Comb-step impact device.

(C) Handrail speed monitoring device.

(D) All of the above.

6. The average speed of escalators in the U.S. is:

(A) 90-100 fpm

(B) 100-110 fpm

(C) 110-120 fpm

(D) 115-125 fpm

7.________ will stop the escalator if its speed exceeds a pre-determined value.

(A) The mainline disconnect

(B) The speed governors

(C) The reversal stop device

(D) The step upthrust device

8. A pallet level device will detect a downward displacement if either side of the trailing edge of a pallet is________.

(A) 1/16 inch (B) 5/16 inch (C) 1/8 inch (D) 3/8 inch

【Extensive Reading】/【拓展阅读】

Special Equipment
特种设备

Special equipment is involved in life, more dangerous boilers, pressure vessels (including gas cylinders, the same below), pressure pipes, elevators, lifting appliances, passenger ropeway, a large recreational facilities, including boilers, pressure vessels (including cylinders), the pressure for the pressure pipe special equipment; elevators, lifting appliances, passenger ropeway, large-scale recreational facilities for the machinery and electronic special equipment.

Special equipment, including its use of the material, attached safety accessories, security devices and protective devices and safety-related facilities.

- **The Boiler 锅炉**

It is the use of fuel, electricity or other energy sources, the costumes, the liquid will be heated to a certain parameter, and exporting thermal energy equipment, the range defined as the volume is greater than or equal to 30L pressure steam boiler; exit pressure is greater than or equal to 0.1MPa (gauge pressure), and the rated power is greater than or equal to 0.1MW of pressure hot water boiler; Organic Heat Carrier Boiler.

- **The Pressure Vessel 压力容器**

It is the dress gas or liquid, carrying a certain pressure in an enclosed device, the range defined as the maximum working pressure greater than or equal to 0.1MPa (gauge pressure), and the pressure and volume is greater than or equal to 2.5MPa.L gas, liquefied gas and the maximum operating temperature greater than or equal to the normal boiling point of liquid container fixed and mobile containers; dressed nominal working pressure greater than or equal to 0.2MPa (gauge pressure), and the product of pressure and volume of large or equal to 1.0MPa.L in the gas, liquefied gas and the normal boiling point equal to or lower than 60 ℃ liquid gas cylinders; chamber and so on.

- **Pressure Pipe 压力管道**

It is the use of a certain pressure, for conveying gas or liquid tubular device, which provides for the maximum working pressure range is greater than or equal to 0.1MPa (gauge pressure) of gas, liquefied gas, vapor or combustible media , explosive, toxic, corrosive, the maximum operating temperature greater than or equal to the normal boiling point of liquid medium, and the nominal pipe diameter greater than 25mm.

- The Elevator 电梯

It is powered, the use of rail running along the stiffness of the box or run along the fixed-line steps (stepping), the lift or the parallel transport of people, goods, mechanical and electrical equipment, including passenger (cargo) elevators, escalators, moving sidewalks, etc.

- Crane Machinery 起重设备

It is used for vertical lifting or moving heavy objects vertically and horizontally of equipment, ranging provisions of the rated lifting capacity greater than or equal to 0.5t lift; rated capacity greater than or equal to 1t, and greater than or equal to 2m lift height and load in the form of a fixed crane hoist and so on.

- The Passenger Ropeway 客运索道

It is powered, the use of flexible rope pulling box to transport personnel and other means of delivery of equipment, including passenger aerial tramway , cable car passengers, passenger towing cable and so on.

- Large-scale Recreational Facilities 大型游乐设施

It is used for business purposes, recreational facilities for carrying passengers, the range provides maximum design speed greater than or equal to run 2m / s, or run a high from the ground than or equal to 2m manned large-scale recreational facilities.

- Special Motor Vehicles within Factory 厂内专用机动车辆

It is in addition to road traffic, farm vehicles just outside the factory in the factory, tourist attractions, playgrounds and other specific areas of special motor vehicle use.

Appendix I

Ref. Translation and Answer

参考译文及习题答案

第1章　电梯结构 I

第1节　电梯概述

电梯是高层商业或住宅建筑物的标准配置。近年来，美国联邦政府残疾人法案要求两层或三层的建筑物也应该加装电梯。

电梯（elevator，英式英语中称为lift）是一种在建筑物不同楼层之间高效运输人或货物的垂直运输设备。电梯一般由电动机或液压泵驱动。在应用垂直输送系统时，一项主要的决策是使用何种驱动系统，液压系统还是曳引系统。每种类型的系统所具备的特性使得他们特别适用于某些场合。一般来说，液压电梯适用于低层建筑（最高6层），曳引电梯（如图1-1所示）适用于高层建筑。

曳引电梯是现在使用最广泛的电梯，由曳引绳和曳引轮之间的曳引力驱动。

电梯本身是比较简单的装置，基本的提升系统在50余年内未发生大的变化。在空间上，电梯可以看成由四部分组成：机房，井道及底坑，轿厢和层站（如图1-2所示）。

无机房电梯（如图1-3所示）是科技进步的结果。新设计的永磁电机（PMM）使得曳引机可以被放置在井道上方，如此一来可以消除在井道上方设置机房的必要。这种设计已经使用了15年，正在成为低层或中低层建筑物的标准产品。最早是由通力引入美国市场。

根据功能来看，电梯由八大系统构成：曳引系统，导向系统，轿厢系统，门系统，重量平衡系统，电力拖动系统，电气控制系统和安全保护系统（如图1-4所示）。

第2节　曳引系统

使用最广泛的电梯设计形式是曳引电梯（或“绳传动电梯”）。曳引电梯比液压电梯更具有通用性，效率也比较高。曳引电梯（如图1-5所示）的轿厢随曳引钢丝绳上下运动，而不是从下面推动。曳引绳与轿厢连接在一起，绕过曳引轮（3）。曳引轮就是一个圆周上有轮槽的定滑轮。曳引轮夹紧曳引绳，所以绳子会随着曳引轮转动而发生移动。

曳引轮与电动机（2）连接在一起。电动机朝一个方向旋转，电梯随曳引轮上升；电动机朝另一个方向旋转，电梯随曳引轮下降。对于无齿曳引机，电动机直接驱动曳引轮旋转。对于齿轮曳引机，电动机通过齿轮组驱动曳引轮。典型的系统是，电动机、曳引轮和控制柜（1）都位于井道上方的机房里面。

提起轿厢的钢丝绳也与对重（4）相连，对重在曳引轮的另一侧。对重的重量等于轿厢及40%的额定载重量。换句话说，轿厢40%载重时，轿厢和对重完全平衡。

平衡的目的是节能。当曳引轮两侧载荷相同，只需要很小的力就能让这个平衡系统朝一个或者另一个方向运动。基本上来说，电动机只需要克服摩擦力——另外一侧的重量起了很大作用。从另一个角度说，平衡保证了系统整体处于一个接近常数的势能等级。轿厢的势能减少（下降到基站），配重的势能增加（上升到井道上方）。当轿厢上升时，刚才的事情逆向发生。系统就像一个跷跷板，每侧有一个重量相等的小孩。

轿厢和对重都沿着井道壁上的导轨（5）运行。导轨防止轿厢前后摆动，而且在紧急情况下协同安全系统制停轿厢。

■ 曳引机和制动器

配备齿轮曳引机（如图1-6所示）的电梯可以用于额定速度达350ft/min（约1.78m/s）的乘客电梯和货物电梯。在少数情况下，它们可在速度高达500ft/min（约2.54m/s）的情形下使用。

无齿曳引机（如图1-7所示）的基础是永磁技术，效率高，节能约60%。它们所具有的低速大扭矩性能，完全取消了齿轮箱，需求空间更小，乘坐完全舒适平稳。

无齿曳引机电梯速度可达2000ft/min（10m/s），或者更高。除了驱动装置不同外，其他零部件与齿轮曳引机电梯相同。

最普通的电梯制动由压缩弹簧总成、有内衬的制动靴和电磁线圈总成构成。当电磁线圈没有通电时，弹簧促使制动靴抱住制动鼓，产生一个制动力矩。磁铁可以施加水平力用于制动释放。这个力可以直接作用在其中一个操作臂上或者通过一个连接系统。制动器与传动轴分开，电梯重新获得速度。

为了改善制停能力，需要在制动器上使用一种高摩擦系数的材料，例如锌黏合石棉。摩擦系数太大的材料可能导致轿厢运行速度突变。这种材料的选择必须慎重。

齿轮曳引电梯和无齿曳引电梯的制动器都是通电释放和断电启用。制动器一般具有鼓的外形，借助弹簧力动作且通电打开，停电会促使制动器动作，防止电梯下降。图1-8对典型电梯制动器进行了说明。

■ 曳引钢丝绳

因为是由多条独立的钢丝构成，所以钢丝绳所呈现的优点让它完全胜任在电梯上的使用。钢丝绳的优点在于它的冗余和能力，识别使用寿命可能终结的能力，或识别在钢丝绳状态恶化前合理时间丢弃的能力，通过外观，如出现钢丝断裂判断。

钢丝绳（如图1-9所示）由多根绳股和一根绳芯构成。中心处的钢丝是一根圆形的钢丝，用作主体。在主体周围螺旋排列的一组钢丝构成了绳股。绳股由绳芯支撑，这样构成了我们所说的钢丝绳直径，在制造中使用的参数。

每股中的钢丝数量越多，钢丝绳弹性越大。每股中的钢丝数越少，钢丝绳越硬。中心可能采用聚丙烯纤维绳芯或者钢质独立钢丝绳芯。

钢丝绳等级和配置众多。对外行人来说，选择钢丝绳的关键因素是断裂强度和直径（测量方式见图1-10）。

充足的安全系数在钢丝绳使用中非常关键。对于杂物梯的曳引钢丝绳，推荐安全系数是10 ∶ 1。换句话说，如果载荷重1t，所用钢丝绳的最小断裂强度是10t。

钢丝绳的直径对于钢丝绳与传动装置的兼容性非常重要。尤其是，槽轮能恰当的容纳钢丝绳以确保它能自由运动而不会产生对钢丝绳或槽轮的过度磨损（如图1-11所示）。

■ 曳引轮/反绳轮/导向轮

曳引轮一般由驱动装置驱动，作用于曳引绳。对于无齿曳引电梯，曳引轮（如图1-12所示）与曳引机传动轴直接相连，电动机的转动（速度）直接传递给曳引轮，没有任何中间传动装置。对于齿轮曳引机电梯，电动机的速度通过蜗杆或斜齿轮减速机构减少至1/10,然后传递给曳引机的曳引轮。

反绳轮是一种定滑轮，一般位于对重架或者轿架,但可能出现在轿厢底部或者井道壁。通过不同的缠绕方式和一定数量的反绳轮可以实现绝大多数的曳引比。

导向轮是一种用于增大轿厢和对重之间距离的滑轮，因为曳引轮的直径有限而某些情况下轿厢的尺寸可能很大。

钢丝绳和滑轮之间存在直接关系，然而这种关系有时被误读。很多时候发生的看上去与钢丝绳相关的问题，如磨损或震动，情理之中的假设就是钢丝绳的制造缺陷。但通常，问题的真正原因在于滑轮。钢丝绳性能不佳是曳引轮问题的表征，因为曳引轮的问题经常会传递给钢丝绳。重要的是牢记钢丝绳位于绳槽，绳槽的形状将会像模具一样传递给钢丝绳直径。

电梯中的曳引力受轮槽形状、材质和表面质量影响，主要的轮槽形状见图1-13。

第3节 重量平衡系统

■ 对重

出现在绳索提升电梯上的第一项改进就是对重。单独提升电梯轿厢需要相当大的工作量，因为上升过程中轿厢重力势能增加。如果能在将轿厢下降时取回储存能量一定很好，可以利用该能量提起对重。

电梯的对重在轿厢上升时下降，在轿厢下降时上升。由于两个物体的质量接近，电梯运行时上升或下降的重量几乎为零。整个电梯的重力势能变化不大，只是在机械的不同部分之间移动。对重平衡了轿厢，所以只需要很小的力量来运转系统。电梯轿厢和对重类似一个平衡的跷跷板，只要一个轻微的推动就可以让它运动。

大多数电梯的对重（如图1-14所示）通过悬挂自身的绳索连接电梯轿厢。绳索从轿厢开始，绕过电梯主轴上方的滑轮，抵达对重。轿厢一般等于电梯空轿厢的质量加40%的电梯额定载荷。这样，当电梯40%满载时，对重与轿厢完全平衡，提升或下移轿厢只需要非常小的工作量。

■ **重量补偿装置**

电梯行程超过100ft（30m）时应具备重量补偿系统。这是一个独立的绳索或链条装置，连接对重的底部和电梯轿厢的底部。如此一来使得控制电梯更容易，因为它补偿了由钢丝绳造成的对重侧和轿厢侧重量差异。

如果电梯轿厢在井道上方，轿厢上方曳引绳长度较短，轿厢下方的补偿绳长度较长，对重正好与之相反。如果补偿系统采用缆绳，在电梯下方的底坑中会增加一个滑轮用于导引缆绳。如果补偿系统采用链条，链条由安装在对重导轨之间的棒材导引。

■ **自我评价**

1. A　2.C　3. A　4.B　5.B　6.D

7. traction system, guide system, car system, door system, weight balance system, electrical drive system, electrical control system, safety protection system.

第2章　电梯结构Ⅱ

第1节　轿厢和门系统

■ **轿厢**

轿厢通过机械方式升降以运送人或货物，在建筑物内从某一层到另一层。对于曳引电梯，轿厢的提升借助于钢丝绳在一个深沟槽滑轮上的缠绕，行业中一般称为曳引轮。轿厢的重量与对重平衡。

轿厢可以分为两大部分：轿厢架和轿厢体。轿厢体（如图2-1所示）由轿底，轿壁和轿顶构成。轿底、导靴、电梯轿厢安全钳、曳引绳或反绳轮及（或）配套设备依附在电梯轿厢架上（如图2-2所示）。

上梁是轿架的上部构件。立柱是轿架的垂直构件，每侧一根，用于锁紧横梁和底梁。

拉杆是一根从轿底延伸到轿厢另一部分轿架的杆，目的是为了支撑轿底或安全地保持在恰当位置。拉杆是轿底外部角落的支撑，每根都与立柱的上半部分联结。

轿底垫块一般是橡胶或其他吸振材料，该吸振材料可以减少传递给轿底的振动或噪声。随着现代化材料的出现，垫块多被新的隔离材料代替，新材料弹性更加利于缓和振动，提升乘客乘坐舒适性。

托架是液压电梯轿架中最下方的水平构件，与压板相连接。曳引电梯轿架最下方的构件底梁装有安全钳。

■ 轿门和层门

当我们讨论电梯门时，轿门和层门是两个不同的概念。在进入电梯前，你面对的是层门；当你处在轿厢内时，你能看到的门是轿门。每台电梯都有一副轿门，但根据建筑物的高度有多套层门。轿门是主动门，层门是被动门。

层门防止乘客坠入井道。图2-3描述了四种常用的电梯层门类型。乘客梯分为中分门，单扇侧开门和双扇侧开门(打开时两扇门重叠)。典型的汽车电梯层门类型是垂直中分门，顶部上升的同时下半部分下沉，从而形成的开口与汽车本身几乎等宽。

一般的描述将电梯门的保护装置分成两类。最早使用而且仍然流行的是机械式保护触板，包括一条轿门前侧的窄带（见图2-4）。窄带在关门过程中向后运动，激活一个开关，导致门停止并反向运动。这种触板采用的材料包括从软橡胶到硬塑料，在某些情况下也可能是金属。由于触板可能在一天内动作数百次，遭受多种多样的接触及滥用，整个机构必须坚固。

另外一种类型是红外保护系统，一种非接触方式（见图2-5）。新式门安全系统采用红外光束覆盖电梯门区域。当光束被隔断时门将缩回。与旧的机械系统不同，新式门不需要与人或物体接触。

第2节　导向系统

■ 导轨和导轨支架

电梯导轨用于保证轿厢或对重上下运动而没有剧烈摇摆，由安装在井道壁上的导轨支架支撑。使用最多的是T形导轨（见图2-6）和用于对重侧或货梯上的空心导轨（如图2-7所示）。导轨在安全钳动作时承受巨大的载荷，所以其硬度或刚性非常重要。

导轨支架（如图2-8所示）的设计是用来将轿厢和对重导轨安全固定在建筑物上。

■ 导靴

导靴用于导引轿厢和对重在导轨上，可以是滑动导靴或者滚动导靴。滑动导靴多数适用于额定速度最高150ft/min（约0.76m/s）。在某些情况下，重载货梯可能使用滑动导靴以获得较高速度。滑动导靴实物见图2-9。

滚动导靴一般用于额定速度150ft/min以上的任何速度。有时用于低速，避免在导轨上进行润滑。滚动导靴见图2-10。

如果导靴维护或调整不良，电梯将会出现摇晃，摆动，发出刺耳噪声或者在井道里一路隆隆作响。

第3节　安全保护系统

每天有数千人乘坐电梯。全球有责任心的业主正在尽自己所能保证人们的安全，通过升级现有垂直输送设备使之符合当地和国家的安全规则。

在好莱坞动作电影的世界中，提升绳索经常会断成两半，导致轿厢及其乘客在井道内直冲而下。实际上，这种事情发生的概率很小。电梯具有多套安全冗余系统确保其位置。

第一道防线就是钢丝绳本身。每根电梯的钢丝绳都是数根钢丝相互缠绕而成。具有这种结构，一根钢丝绳本身就可以承载轿厢和对重的重量。但电梯一般具有多根钢丝绳，典型的是4 ~ 8根。即便意外情况下一根钢丝绳断裂，其余钢丝绳会吊住电梯。

即使所有钢丝绳都断裂或者与曳引系统脱离，电梯轿厢也不太可能坠落到井道底部。曳引电梯轿厢具有内置制动系统或者安全钳，这是一种在轿厢运行过快时将其在导轨上夹牢的装置。

▪ 限速器

限速器安全系统是一种独立的安全系统，目的并不是调整速度或者在正常运行时制停电梯。轿厢的运动传递给限速器，限速器一般不会动作，除非轿厢超速。

当轿厢或对重运动时，限速器绳带动限速器旋转。在某一设定速度，限速器将激活一个开关，对驱动主机和制动器断电。如果该操作未能阻止电梯下降，限速器卡住限速绳，导致安全钳动作使得电梯停止。典型限速器见图2–11。

▪ 安全钳

安全钳是固定在电梯轿厢上的制动系统，在井道内随电梯上下，紧急情况下夹住导轨。一部分安全钳是夹住导轨，而其他的是将楔形块压入导轨缺口。有代表性的是，安全钳由机械式限速器触发。

安全钳座利用横向拉杆或类似零件安装在轿架上。为了避免电梯轿厢横向移动时意外咬住楔块，允许安全钳底座与杆一起横向移动。

安全钳与导向装置在每侧导轨处放置，以便导向装置与导轨的间隙小于楔块与导轨的间隙。因此，当电梯轿厢架相对于导轨的位移超过导向装置和导轨之间的距离，安全钳底座横向移动，保证了导轨和楔块之间的最小间隙。

安全钳可以分成两类：瞬时安全钳和渐进式安全钳（如图2–12所示），前者用于低速电梯，后者用于高速电梯。

▪ 夹绳器

最初的限速器和安全钳可以阻止轿厢墩底，但是不能阻止轿厢冲顶。

A17.1—2000法规要求，轿厢上行保护和意外运动防护。对于老旧电梯可以通过在钢丝绳上单独采用一个制动器或者在驱动轮上施加一个制动器来实现。为了防止轿厢冲顶，现在有另外一种装置称为夹绳器，见图2-13。

注释：ASME A17.1—2000是《电梯和自动扶梯的安全规范》(SAFETY CODE FOR ELEVATORS AND ESCALATORS)是美国的国家标准。

■ 缓冲器

缓冲器是电梯安全系统的重要一部分，是电梯失控的最后保障。它可以很大程度上降低或缓和电梯在紧急情况下墩底的冲击力，保护乘客的安全和货物免受毁坏。

有三种类型的缓冲器应用于不同的电梯，分别是弹簧缓冲器、聚氨酯缓冲器和液压缓冲器（如图2-14所示）。

弹簧缓冲器是缓冲器的一种，常见于液压电梯或用于速度低于200ft/min的曳引电梯、货梯、杂物梯或者仅在对重侧。该装置用于缓和电梯冲击，常见于电梯底坑。

液压缓冲器是缓冲器的另一种，常见于速度高于200ft/min的曳引电梯。该类型的缓冲器综合利用油和弹簧缓和下降的轿厢或对重，所以这是一种耗能型电梯缓冲器。

缓冲器因位于底坑而容易暴露于洪水中或进水。需要对其例行清洁及喷漆，确保维持正常的性能。油压缓冲器需要进行油位确认，暴露于洪水后需要换油。

第4节　电气控制系统

自动控制的电梯最早出现于20世纪30年代，因纽约和芝加哥等大城市日益增多的摩天大楼而加速发展。采用继电器回路的机电系统来控制电梯的速度，位置和门的动作。继电器控制电梯在20世纪80年代以前非常普遍，逐渐被以微机为基础的控制设备所代替，而微机为基础的控制设备已成为当今工业标准。

■ 轿厢操纵面板

操纵面板（如图2-15所示）上装有电梯操作所必须的各种元器件。采用不同的按钮系统可以实现范围很大的设计。面板一般采用镜面不锈钢，如果需要也可以采用发纹不锈钢。

层数比较多时建议采用双排的轿厢操纵面板，因为单排的会太长。双排面板的长度比单排面板的小，以便所有人都可以碰到按钮。

■ 控制柜

控制柜（如图2-16所示）对于电梯来说非常重要，类似于人类的大脑，不断地从轿厢操纵面板或召唤盒接收信号，给曳引机发出动作指令。

大多数电梯的控制柜安装在机房。无机房电梯的控制柜安装在井道内。

现在出现了具有“目的地控制系统”功能的电梯，乘客在进入电梯前通过目的地操作面板登记目的楼层。尽管行程时间缩短了，但乘客等待时间可能会更长一些。

■ 召唤盒

召唤盒（如图2-17所示）或者厅门召唤盒在每个层站都有。大多数层站的召唤盒至少有两个上行和下行的召唤按钮，端站除外。召唤盒的结构与轿厢操纵面板非常类似。

■ 自我评价

1. D 2. B 3. B 4. A 5. A 6. car frame car body 7. contactless 8. roller 9. the former（the instantaneous type）

第3章 金属基本知识

第1节 金属分类

在现代工业中，有近千种金属应用于生产。现代汽车的制造就需要用一百多种金属。下面，我们将试图通过这篇文章让大家对金属的基本分类有个了解。

金属过去被认为是具有金属光泽、良好的导电性和导热性的物质。实际上，金属一般被定义为其氢氧化物为碱性的物质（如钠、钾），而非金属则定义为其氢氧化物为酸性的物质（如硫黄）。金属可以是由单一元素构成的纯金属，也可以是由两种或两种以上金属元素组成的化合物，这种化合物就称为合金。

“合金”这个术语用于识别任何金属系统。在冶金学中，合金是由两种或两种以上的元素均匀混合、具有金属特性的物质。在所构成的元素中，必须有一种是金属。例如，普通碳钢就是主要由碳和铁组成的合金，当然还包括其他的一些杂质元素。然而，出于商业目的，普通碳钢没有被归类为合金。

纯金属有固定的结晶温度，但合金没有。合金又可细分为铁合金和非铁合金，其区别就在于是否包含铁元素。

所有的商业用钢、铁都是合金。最普通的钢就是铁和碳的合金，但实际上，它还包含微量的硅元素和锰元素。另外，还有很多的合金钢，例如特殊工具钢（用于铸造、锻造和辊型）等，这些合金的基底元素都是铁。

通常可根据合金中的主加元素来区分钢的种类，例如硅钢、锰钢、镍钢和钨钢等。即使在非铁合金中铁也可以作为杂质少量地存在，如青铜、黄铜和蒙乃尔铜。

非铁金属在自然界很少以纯金属的形态存在。要想获得纯的金属，就要从矿石中将脉石分离出来。在分离之前要先选择合适的矿石，这个过程通常称为选矿。纯金属或者金属化合物要比脉石重，当我们将矿石在热水中搅拌后，纯金属和金属化合物将会沉淀于底部被分离出来。这种方法有点类似于早期矿工的淘金。然而，利用这个原理发展起来的精炼技术提高了金属化合物的凝聚速度。

非铁金属的熔化最常用到反射炉。此种炉的构造包含两层，里层由耐火砖组成，外层是钢。原料放入其中，间接进行加热，在加热过程中，要在原料中加入一些诱导物，造渣物或熔剂，用于脱氧。

第2节　金属特性与热处理

金属具有很多区别于其他材料的特性。其中最重要的就是强度，它是一种承受载荷不发生弹性变形和折断的能力。这种性质非常重要，它通常是和韧性——金属发生变形而不断裂的能力联系在一起的。除此之外，金属还具有抗腐蚀的能力，这种能力的高低与热处理相关。

金属可以铸成许多不同的形状和尺寸。它们既可以进行焊接处理，也可以进行淬火和退火处理。回收再利用是金属的另一个很重要的性质。当某一产品不能再被使用时，我们就可以将它分割成方便处理的若干小块，然后将它们放进炉子内，重新熔化，铸成新的产品。

金属的特性可以分为三类：即化学特性、机械特性和物理特性。这里我们着重介绍金属主要的机械特性。在了解金属加工的相关领域以及目前所采用的加工方法时，金属机械特性就显得尤其重要。

金属硬度的变化范围很大。有些金属，例如铅，很容易塑造成形。还有些金属，例如碳化钨（硬质合金），它的硬度已经接近钻石的硬度。它们在制造各种硬质合金刀具时具有很高的价值。热处理可以导致金属硬度发生变化。退火工具钢易于机加工，但经淬火和回火（调质处理）后，机加工就变得困难了。退火以后的黄铜相对来说较软，当把它进行冷加工以后，它的硬度则大大提高。

很多世纪以前，人们就已经知道了钢具有在高温条件下骤然冷却可以提高硬度的特性。对金属做这种处理就称为热处理。它是冶金学的一部分，通过加热改变了金属的结构。

淬火：在任何一种热处理的操作中，加热的速度都是非常重要的。热量从钢的外部以一个确定的最大速度向内部传递，如果加热的速度太快，零件的外部温度就会高于其内部温度，这样将会很难获得内外均匀一致的组织结构。

通过热处理可获得的硬度主要与以下三方面的因素有关：淬火速度，碳的含量，工件的尺寸。

普通低碳钢和中碳钢为了淬硬，应该采用快速淬火工艺，通常采用水作冷却介质。对高碳钢和合金钢，则采用油冷，油冷的淬硬作用比不上水冷。如果要求严格冷却，就必须得用盐水了。

直接冷却后的钢所能获得的最大硬度在很大程度上由碳的含量所决定，因此，低碳钢即使在热处理后也不可能达到太高的硬度。碳钢一般也被称作浅硬化钢。不同钢的淬火温度是不一样的，这是由碳的含量所决定的。

通常使钢淬火变硬的温度，我们称为淬火温度。它通常要高于加热转变临界温度10 ~ 38℃，在这个温度时金属的组织结构就会发生变化。

回火：淬火使高碳钢和工具钢变得极其硬而且脆，大多数情况下不能直接使用。通过回火，淬火过程中产生的内应力得以消除。回火提高了淬火零件的韧性，也使材料具有更大的塑性或延展性。

退火：退火是将金属稍微加热到临界温度以上后，很缓慢地冷却。退火处理能够减轻金属内部由于先前热处理、切削加工或其他冷加工所造成的内应力和应变。钢的种类决定着退火加热的温度，加热的温度也与退火的目的有关。

完全退火用来最大限度地降低钢的硬度，以改善它的切削加工性能，消除内应力。低温退火也叫做去应力退火，它的目的主要就是消除在冷加工和机械加工过程中产生的内应力。球化退火是使钢中生成一种特殊的晶粒结构，这种结构相对较软而易于加工。这种工艺一般用于改善高碳钢的切削加工性能和用于拉丝工艺的热处理。

正火：正火操作是用来消除金属由于热加工、冷加工及机械加工过程中产生的内应力的过程。正火是把钢加热到临界温度以上30 ~ 50℃，保温一段时间后空冷。正火通常应用于低、中碳钢和合金钢。正火可以消除先前热处理所留下的各种影响。

第3节　金属加工机床

机床的工作原理与一般工具的工作原理相同。一般来讲，机床的使用目的就在于通过改变毛坯的形状、大小或材料表面光洁度等，使之成为机器零件。机床加工的工件形状是有限的，其形状主要取决于刀具运动的类型。为了控制刀具的运动，就必须设计许多不同种类的机床。

标准机床通常可划分为以下五个基本类：车床，钻床，刨床，铣床和磨床。每一种机床都可以改变工件的形状、尺寸及表面粗糙度等。这种加工可以通过工件上的材料来实现。每一种机床都根据这样的原理来工作：对刀具或者工件施以运动并使两者相接触。当机床与工件联系在一起时，所有机床的操作都遵循最基本的规则。然而，每种机床在所采用的运动方式或所应用的方法上，又各不相同。下面针对每种机床的基本操作特性作以简单的介绍。

车床是一种使用刀具边缘部分加工回转体零件的机床，刀具切削路线是曲线，但刀具的进给路线在大多数情况下是直线。加工出来的工件基本上是圆柱体。

钻床通过多刃刀具的旋转来产生所需的切削运动。旋转的刀具垂直于工件的表面进行连续的切削。多刃刀具旋转与直线进给两种运动合成后，就可以在工件上加工出满足要求的孔。

仿形机床和刨床可以放在一起进行说明。两种机床都使用点接触刀具加工平面。刀具和工件所走的路线基本相同，都是往复的直线运动，往复运动指的是前后移动。刀具的形状、作用等都是类似的。最主要的不同在于仿形机床的工件是固定不动的，它是通过刀具的往复运动来完成加工的。而对于刨床，则是工件进行往复的运动。通过刀具的往复运动来实现连续的切割。在每一个行程中，工作台横向移动，刨刀则以一定的进给量自动进给。

铣床对旋转的刀具提供切削运动，工件沿直线进行往复运动，旋转的刀具直线进给进行加工。经铣床加工后的表面一般来说是沿某一直线方向的轨迹。铣床所使用的刀具是多刀刃的。加工的过程就是通过这些刀刃的边缘部分来完成的。

磨床所能达到的加工精度是所有机加工方法中最高的。它与其他机床的区别在于它的刀具（砂轮）是由金刚砂、碳化硅等类似原料制成的。砂轮的表面由许多极小的微刃（砂粒）组成，操作者通过移动工件与砂轮表面接触。在砂轮的高速运动中，以除去工件材料微粒的方式完成切削。磨床的加工运动除了控制砂轮的旋转外，还可以通过控制工件的旋转运动和往复运动与砂轮的旋转方向（或所有这些运动结合起来），来完成连续的加工。

机床的应用大大地推动了现代文明的发展，已广泛应用于汽车、铁路、飞机、轮船甚至宇宙飞船等领域。如今的大批量生产技术归功于现代机床的应用，这些技术使得人们获得了更多的产品，而这些产品已经融入了人们的生活。

- **自我评价**

1. B　2. A　3. A,C,B　4. C　5. D　6. 略

7. hydroxide, metal, strength, toughness, water, normalizing, lathe

第4章　钣金加工

第1节　钣金成形

钣金成形是通过施加外力改变其形状，而不是移除材料。

施加的外力超过材料的屈服强度，造成材料发生塑性变形，但不是失效。通过这个过程，钣金可以被折弯或拉伸成各种各样的复杂形状。

将工件归为钣金的材料厚度并没有明确定义。然而，一般认为0.006 ~ 0.25in厚度的原材料为钣金。更薄的称为箔，更厚的称为板。

- **折弯**

弯曲成形是施力使金属板弯曲成一定的角度并形成期望的形状金属板成形过程。弯曲操作可使金属板沿某一轴线变形，而执行一系列不同的操作就可以形成复杂的零件。

弯曲零件可小到支架，大到20ft（约6m）的外壳和底座。折弯可以通过几个不同参数来表征，见图4-1。

弯曲的动作使金属板产生了伸张和压缩。金属板外部受到拉伸而产生伸展，

内部受到挤压而缩短。中轴线是金属板的分界线，中轴线上既无伸张力也没有压缩力存在。因此，这个线的长度保持恒定。

金属板内外表面长度的变化与最初平板长度有关，由两个参数决定：弯曲余量和折弯补偿，定义见图4−2。

当弯曲钣金时，金属本身的残余应力会使得板在弯曲操作后轻微的回弹。正因为存在弹性恢复，所以有必要将金属板过度弯曲到某个精确的量，以达到期望的弯曲半径和弯曲角度。最终的弯曲半径要比最初形成的大些，而最终的弯曲角度要比最初形成的小些。最终弯曲角与最初弯曲角的比率被定义为回弹系数K_s。决定回弹量的因素很多，如材料自身性质，弯曲操作方式、最初的弯曲角度和弯曲半径。

折弯通常是通过弯板机实现的，既可以手动也可以自动。因此，折弯有时也称为弯板成形。弯板机规格从20 ~ 200t不等，以满足不同的应用场合。

弯板机（如图4−3所示）包括上半部的冲头和下半部的模具，钣金在中间。当冲头下降并折弯板材时，钣金需要通过后座等位置进行良好的定位。在自动折弯机中，冲头被液压连杆带动冲压钣金。弯曲角度通过压入模具的深度来决定，深度是精确控制的以达到预期的折弯。

标准工具常用于冲头和模具，初始成本低并满足小批量生产。专用工具用于特殊的操作，但是会增加成本。

工具的材料取决于产品的数量，钣金的材质，弯曲的角度。好的工具需要承受数量多、强度大、苛刻的弯曲。为了提高强度，可以采用低碳钢、工具钢和合金钢。

即便采用弯板机和标准冲压零件，还是存在许多折弯技术要点。最常采用的方式为V形折弯，冲头和模具的形状是V形的。冲头将钣金推进V形槽使其弯折。

如果冲头没有把钣金推到V形槽底部，称为空折（如图4−4所示）。这时V形槽的角度比需要成形的钣金角度更尖。如果推到底部，称为底折。这时要考虑角度控制，因为回弹较少。回弹值称为模具比率，等于模具开口值，只是被材料厚度隔开而已。

除了V形弯曲之外，另外一种常见的折弯方式是擦弯曲（如图4−5所示），有时称作边界弯曲。擦弯曲要求钣金被一个压头压在擦头上。冲头贴着边进行施力，钣金沿着擦头进行折弯。

一些设计准则列出如下：

1) 折弯位置——折弯应该位于材料足够多的地方，因为需要确保钣金没有滑移，所以最好是直边。孔和槽在内的任何特征如果靠近折弯位置太近会出现变形。这类特征距离折弯位置的距离应大于3倍的材料厚度，不包括折弯半径。

2) 折弯半径——所有折弯采用相同半径消除额外的模具或机构。折弯内径至少为一个料厚，折弯内径过小可能导致材料折弯时破裂。

3) 折弯方向——硬度较高的金属沿着滚扎方向折弯时，钣金容易出现破裂，建议折弯垂直于滚轧方向。如果采用手工折弯，在设计允许的情况下，可以沿折弯位置切出一条槽，以此来降低手工折弯时需要的力。

■ 深拉成形

深拉成形零件的特征是深度大于折弯直径的一半。这些零件的截面可以多种多样，笔直的、锥形的，甚至曲线形状，但圆柱形和长方形的零件是最常见的。

深拉成形（如图4-6所示）尤其适用于易延展金属，例如铝、黄铜、红铜和低碳钢。深拉成形的零件例子包括汽车车身、油箱、金属罐、杯子、厨房水盆和炊事用具。

第2节 钣金切割

切除过程是通过施加一个足够大的外力，使材料分离的一种失效方式。最常见的切割过程是施加剪切力，因此有时也称为剪切过程。当施加一个足够大的剪切力时，材料内的剪切应力超出剪切强度，这时材料就从切割处失效并分离。

■ 剪切

存在各种各样利用剪切力的切割过程，以不同的方式从母板片料中分离或移除材料。然而，剪切这个词本身指的是一个具体的剪切过程，产生切割线来分离金属板。因为这个原因，剪切（如图4-7所示）一般用来将母材切割成小块做预加工。

剪切力通过两个刀具施加，一个在钣金上方，一个在钣金下方。无论刀具称为冲头和模具或是上刀模和下刀模，上方的刀具快速向下移动到靠在下方刀具的钣金上。上下刀具之间存在很小的间隙，用于撕裂材料。间隙大小一般为材料厚度的2% ~ 10%，取决于下述几个因素，例如具体的剪切过程、材料和钣金厚度。

剪切过程通过剪切设备实现，一般称为平行刃口剪切机或者动力剪，操作可以通过手动（手或者脚）、液压、气压或者电力来实现。 一部典型的剪切设备包含一个具有支撑壁的工作台来放置金属板、限位块、金属板导引、上下刀头，准确定位板材的定位装置。 金属板放在上下刀具之间，接下来上下刀具会在力的作用下一起挤压金属板，切除材料。在多数设备中，下刀具静止而上刀具运动。

■ 下料

下料（如图4-8所示）是这样一种切断方式，通过施加一个足够大的剪切力，从一大片母材中移除一块材料。在这个过程中，被移走的称为坯料，不是废料而是想要保留的部分。下料件一般需要二次加工以去除边缘的毛刺。

下料过程需要一个下料压头，金属板母材，下料冲头和下料刀模。钣金原材料放在型腔上方。型腔不是凹坑而是一个与预期零件相同形状的切口，除标准形状外必须定做。在钣金上方，设有下料冲头，与期望零件相同形状的刀具。

精下料（如图4-9所示）是一种特殊类型的下料方式，通过施加3种独立的力从母材下料。该技术生产的零件平面度更好，边缘光滑毛刺少，误差可达

±0.001mm。如此一来，不要二次加工就可以下料得到高质量的零件。然而，额外的设备和刀具增加了成本，使得精下料更适于大批量生产。

■ 冲裁

冲裁是一种通过施加足够大剪切力从一片母材上移除材料的剪切过程。冲裁与下料非常类似，只是称为金属块的移除材料是废料，而在片材上留下所期望的内部特征，如孔或者槽。

冲裁（如图4-10所示）可以用于生成不同形状和尺寸的切口和孔洞。最常见的冲孔形状简单，如圆、方形和长方形等。冲孔特征的边缘在被剪下时会有一些毛刺，但质量还较好。二次精加工操作一般是为了获得更光滑的边界。

■ 自我评价

1. A　2. C　3. B　4. C　5. B　6. A　7. C　8. B　9.B　10. C

第5章　焊接技术

第1节　熔焊

■ 电弧焊

电弧焊是一种利用焊接电源在电极和基材之间产生电弧，从而在焊接点熔化金属的焊接类型。可以使用直流电（DC）或者交流电（AC）。

这些方法中应用广泛的是焊条电弧焊（SMAW），也称为手工电弧焊（MMA）（如图5-1所示）。它定义为一种通过在有覆盖层的金属电极和工件之间加热产生熔合的电弧焊。防护通过电极覆盖层的分解获得。不需要压力，填充金属通过电极获得。

该焊接方法用于多个领域，尤其是机械设备生产制造、运输设备和管道系统。

埋弧焊（SAW）（如图5-2所示）是一种通过在裸露金属电极和工件之间加热产生熔合的电弧焊。电弧受到工件上一些颗粒状易熔物体的保护。不采用压力，填充金属通过电极获得，有时也来自于补充焊条。

该方法可以用于电极和颗粒助焊剂均可以控制的全自动设备。该方法也适用于电极和颗粒助焊剂手动控制的半自动设备。因为颗粒助焊剂必须覆盖在焊接点上，采用该方法时零件必须水平放置，尤其适合长直线型接缝。此外，焊接较厚金属时比焊条电弧焊所需要的次数要少。

等离子弧焊（PAW）（如图5-3所示）是一种通过电极和工件之间的约束电弧加热产生熔合的电弧焊。防护通过孔口散发的热电离气体获得，亦可能通过辅助防护气体进行补充。防护气体可能是惰性气体或混合气体。

■ 气焊

气焊是一种火焰通过压缩气体燃料供给的焊接。根据使用场合、获得可能性和成本，可以使用几种不同的气体燃料。气焊是一种危险的焊接方式，因为使用的是易燃易爆的气体，而且储存在压力容器中。气焊从19世纪开始就被积极利用，尽管在很多应用场合已经很大程度上被电弧焊取代，但因其携带方便仍然在使用。

氧－乙炔焊接（OAW）（如图5-4所示）是使用最多的一种，一般称为氧焊或者在美国称为气焊。该方法采用氧气和乙炔混合产生热量。采用助焊剂可减少氧化并提高焊接质量。该焊接方法适用于铁质（包括铸铁）和非铁金属质材料，可用于焊接厚截面金属。

氧－乙炔焊接的主要优点在于设备简单，便于携带，成本低廉。所以，在维修和维护时非常方便。然而，由于其功率密度有限，焊接速度非常慢，单位长度的热量相当大，导致热影响区很大及严重变形。氧－乙炔焊接不适于焊接钛和锆之类活泼金属。

氢氧焊（OHW）用于低熔点金属，例如铝、镁和铅。气压焊（PGW）利用氧乙炔火焰作为热源，但不需要填充棒。该焊接形式能用于焊接铁质和非铁金属。

第2节　压力焊

■ 电阻焊

电阻焊通过电流经过电阻产生热量，电阻产生在两个或多个金属表面之间。当强电流（1000 ~ 100000A）流经金属时，在焊接区域产生小的熔池。总的来说，电阻焊效率高，产生轻微污染，应用在某种程度上受限，成本较高。

点焊（如图5-5所示）是一种普遍的电阻焊，用来焊接重叠在一起厚度达3mm的钣金。两个电极同时夹住金属板，以便电流经过金属板。这种方法的优点包括能量使用效率高，工件变形有限，生产效率高，易于自动化和无需使用填充材料。

点焊强度比其他方法低，使得该方法只适用于特定场合。点焊在汽车工业中应用广泛，普通汽车通过工业机器人有数千处点焊。一种称为瞬时点焊的特殊制程，可以用于点焊不锈钢。

与点焊类似，缝焊（如图5-6所示）依靠在两个电极施加压力和电流来熔接金属板材。然而，并非点状电极，而是采用轮状电极沿着工件滚动并实现工件进给，使得进行长而连续的焊接变得可能。

在过去，该过程用于饮料罐的制作，但现在使用有限。其他的电阻焊方法包括对接焊、闪光焊、凸焊和电阻对接焊。

第3节　钎焊

钎焊一直以来是熔接金属的最广泛最有效方法之一，直到近些年，由于其设备庞大而且昂贵，因此少有使用。

钎焊是一种金属连接过程，填充金属在上方被加热后，通过毛细管作用分布在两个或多个紧密相连的零件之间。填充金属被加热到稍高于其熔化温度，同时利用助焊剂进行合理的气体防护。然后流经基底金属，接着冷却，与工件熔接在一起。

钎焊类似于锡焊，但是熔化填充金属的温度高于450℃，或者是427℃（800℉，美国的传统定义）。钎焊在没有熔化零件和无零件变形的情况下产生强度极高的连接点，连接点的强度一般大于基材本身。

■ 火焰钎焊

火焰钎焊是目前为止在用机械式钎焊的最常见方式。它最适于小批量生产或专业化经营，在某些国家它占了钎焊的大部分。火焰钎焊需要使用助焊剂防止氧化。有三类主要的火焰焊接：手工方式、机器方式和自动化方式。

手工火焰钎焊是一种在焊接结点上方或附近通过气体火焰产生热量的过程。焊枪可以手持或者固定位置，取决于操作是否完全手工或者具有一定的自动化程度。手工火焰钎焊一般用于小批量生产或者零件尺寸形状特殊无法利用其他钎焊的场合。主要缺点在于方法及其获得高质量钎焊结点的操作工技能所需的劳动成本高。

机械化火焰钎焊一般用于需要完成重复钎焊操作的情形。该方法混合了自动和手动操作，操作人员放置钎焊材料，助焊剂并用夹紧零件，机器装置完成实际钎焊过程。该方法的优点在于降低了手工火焰钎焊对技能的要求和劳动强度。自动化火焰钎焊是一种基本消除了人工操作的钎焊方法。

■ 炉内钎焊

因为炉内钎焊（如图5-7所示）适于批量生产且不需要熟练工人，使其成为工业钎焊中使用广泛的半自动制程。相对于其他加热方式有众多优点，使得炉内钎焊成为批量生产的理想方式。

优点之一是生产大量小零件时夹紧简单或易于自动定位。该过程还提供了受控热循环的好处，而且无需钎焊后清洁。其他优点包括：批量生产时单件成本低，封闭温度控制，一次钎焊多个结点。炉内钎焊一般采用电力、燃气或燃油加热，取决于焊炉类型和应用场合。

然而，该方法的缺点包括：设备投资成本高，设计考量困难以及高能耗。

第4节　焊接质量控制

确保焊接工人遵守具体的程序是全部焊接质量系统中关键的一步。许多明显的因素影响焊接点和周围材料的强度，包括焊接方法、输入能量的数值及集中度、基材的焊接性能、填充剂、助焊剂、结点设计（如图5-8所示）和上述因素的相互作用。

焊接缺陷的类型包括裂纹、变形、气孔、非金属夹杂物、未熔合、未焊透、层状撕裂和咬边（如图5-9所示）。为了确认焊接质量，经常采用破坏性测试方式及无损检测方式确认焊接是否存在缺陷，残余应力和变形是否处于可接受水准，热影响区（HAZ）特性是否可以接受。

■ 无损检测

无损检测是一种不破坏或损伤焊接件使用性能的测试方式。这些测试能揭示常见的因使用不当操作步骤而引发的内部和表面缺陷。有众多测试仪器可供选择，其中的大多数比使用破坏性测试方法的还要简单，尤其是对于大型和昂贵的物件。

目视一般由焊接工人完成，焊接后自行检查。这是一项完全主观的检查类型，通常没有明确的或严格的接受界限。焊接工人可以使用样板来确认焊缝轮廓。该测试只有在焊接外观质量最重要时有效，否则只能作为参考。

磁粉检测对于表面和接近表面焊接缺陷的探测是最有效的。用于能感应磁化的金属或合金。测试品被磁化后，涂抹磨得很细的铁粉。只要磁场没有中断，铁粒会在测试品表面形成规则模式。当磁场被金属裂纹或其他缺陷打断时，悬浮的金属也会被打断。金属簇的颗粒出现在缺陷周围，使得缺陷位置很容易确定。

如图5-10（a）所示，电流流经测试品并磁化它，或者如图5-10（b）所示通过给缠绕在测试品上线圈通电磁化。当电流沿直线从一接触点到另一接触点时，磁力线是圆周方向的，如图5-10（a）。当电流经过测试品周围的线圈，磁力线沿长度方向，如图5-10（b）。

液体渗透方法用于检查金属表面缺陷，类似于磁粉检查方法。与磁粉检测可以揭示表面以下缺陷不同的是，液体检测只能揭示开放表面的缺陷。

射线检测是一种通过射线穿透焊缝检查焊接件的方法。X射线或伽马射线是该制程采用的两种光波。射线经过焊点到达与焊点背后直接接触的感光胶卷。胶卷冲洗后，气孔、夹渣、裂纹或未焊透会呈现在胶卷上。因为射线具有危险性，只授权给有资质的人员进行测试。

超声波检测利用高频振动或声波定位及测量焊接缺陷。它既可以用于铁质材料，也可以用于非铁金属材料。这是一种非常灵敏的系统，可以定位细小的表面和内部裂纹，以及其他类型的缺陷，可以测试所有结点类型。该制程采用高频脉冲确认焊接的完好性。在良好的焊接件中，信号从一端经过焊接点到另一端，接着反射回来，显示在一个校准过的屏幕上。例如气孔或夹渣等不规则情形会导致信号提前返回，以深度变化显示在屏幕上。使用该系统可以对绝大多数材料进行缺陷确认。

涡流检测是另外一种使用电磁能量发现焊缝熔敷缺陷的测试类型，对铁质和非铁金属材料均有效。涡流检测原理为，当靠近金属的线圈内有变化的高频电流时，通过感应在金属内产生电流，该感应电流成为涡流。焊接组织的不同导致线圈中阻抗的变化，反应在电子设备上。当存在缺陷时，显示阻抗变化，缺陷大小通过变化值显示。

■ 破坏性测试

在破坏性测试中，需要焊接结构的样品。这些样品承受载荷直到真正的失效。接下来研究失效品并与已知标准比较，以确定焊接质量的好坏。破坏性测试的主要缺点在于，必须用实际的焊接件切片来评估焊接。该类测试方法一般用于焊接工人制程的确认。

切口测试对于确定焊接内部质量非常有用。该测试揭示各种内部缺陷，例如夹渣、气孔、未熔合，以及金属氧化或烧伤。为了完成对焊的切口测试，必须先利用火焰切割从焊接样品上切下样本。在两侧通过焊接中心的位置各制作一个切口。接下来，将锯过切口的样品放在两个钢铁支撑块上面，如图5-11所示。在制作锯槽的位置，利用锤子砸断样品。

自由弯曲测试设计（如图5-12所示）用于测量焊缝熔敷和焊接相邻热影响的延展性。它也用来确定焊接金属的延伸百分率。你应该记得延展性是金属允许被拉长或锤薄的特性。

术语抗拉强度可以定义为截面上用磅每平方英尺来衡量的纵向应力和拉伸抵抗能力。抗拉强度测试（如图5-13所示）包括在强度测试机器上放置试品以及拉伸测试品直到断裂。

■ 自我评价

1. B　2. B　3. D　4. B　5. A　6. A　7. B　8. A　9. A　10. D

第6章　用户指南

第1节　曳引机规格书

与博达技术上领先一步的理念一致，博达已经开发出一款新的型号为BETM-06A的机器，其载客量达6人的低成本解决方案能满足顾客在这方面的需求。

BETM-06A型机器（如图6-1所示）配有改进型的电动机，以提供无颠簸的平稳行进。所有关键机器零部件均产自博达自有的制造工厂，以获得预期质量。每台机器都经过博达最新测量和测试设备的严格检验，以确保每台机器符合设计标准。

可以根据客户要求，提供有底座型和无底座型BETM-06A机器。

■ 显著特征

- 电动机使用电压范围广。
- 同一机器可以用于直接启动和交流调频调压应用。
- 电力驱动装置（定子和转子）集成安装在高速主轴上。
- 尺寸紧凑重量轻。

- 整体底座材质为灰铸铁。
- 曳引轮材质为球墨铸铁。
- 高速主轴为硬质合金制造。
- 蜗轮以特级磷青铜为原料，以获得更长的寿命。
- 制动器动作为电磁原理，在110V直流电或220V交流电下双磁头动作。
- 适于使用矿物润滑油。
- 齿轮和轴承通过油池飞溅润滑。
- 易于在工作现场维护。

■ 规格参数

- 最大载荷能力：6人或408kg。
- 静态载荷能力：无底座为1725kg，有底座为2375kg。
- 额定最大速度：0.63m/s。
- 齿轮箱输出扭矩能力：921N·m（输入速度940r/min）。
- 电动机功率及速度：3.7kW/940r/min，380V/3相/50Hz，绝缘等级F。
- 制动器动作原理：直流电110V电磁型。
- 曳引轮尺寸：直径530mm×3绳槽×钢丝绳直径10mm或13mm。
- 净重（无油）：225kg。
- 油量：3.5L。
- 减速比：41：1。

第2节　曳引机使用手册

■ 1. 简介

祝贺你选择了博达曳引机。本手册提供该曳引机的安装、操作和维护说明。请在开始使用产品时从头到尾看一遍这本安装、操作和维护手册。

机器良好的工作取决于精心的安装，正确等级的润滑油和良好的工作条件。因此，确保机器根据手册给定说明进行安装非常重要，以确保机器工作良好和长时间无故障运行。请注意如下事项：

- 手册中描述的操作必须由配备适当工具和设备的有资质人员完成。
- 在执行任何维护工作前，系统必须停止运行。
- 订购备品时必须按照惯例引用曳引机的序列号。
- 所有曳引机均包装在木箱或板条箱中。
- 曳引机必须使用适当的设备从卡车上仔细卸载。
- 收货时应确认机器状况。如果机器损坏，没有博达许可/认可请不要安装。
- 曳引机应在一干燥场所储存在原始包装中，免受天气影响。
- 如果曳引机包装已经打开，必须防止灰尘。
- 客户应该采用安全护栏对曳引轮和旋转轴扩展部分进行防护。

▪ 2. 长期储存的影响

如果曳引机搁置不用或长时间未投入运转（从发货起超过12个月），机器性能很有可能受到影响，在这种情况下博达不能保证安装后有良好的性能。建议将机器整体送回博达工厂确认轴承、油封、橡胶件和机器的状况。

▪ 3. 安装与操作

3.1 搬运

- 曳引机供货时组装完好且无润滑油。曳引轮涂有防锈剂，只能通过合适的溶剂清除。任何情况下不得通过刮除和锉掉等方式清除防锈剂。
- 曳引机应通过吊环螺栓或者整体式铸造凸耳吊装。这意味着曳引机只能单独吊装，任何配件不得与曳引机一同吊装。
- 曳引机可以使用链条或皮带吊装，但必须留意以避免在关键零件处施加载荷。

3.2 安装

- 将曳引机放置在底座并确保其安放在所有接触点上，也包括如果存在的外部支撑与固定孔位相一致。如果接触不平整，保留部分包装以确保底座完全放置在地板上。
- 按照对角顺序拧紧螺栓，将曳引机固定在底座上。
- 确保地基具有足够的承载能力。

3.3 润滑

- 润滑油采用矿物油。
- 日常的油位确认非常有必要。
- 加油说明：在曳引机停止运转时，将油从孔部注入，到达球形油位指示器中心线为止。
- 排油说明：在曳引机停止运转时，拧开底座上的放油塞，让全部油流干。

重要预防措施

✧ 首次换油应在运转5000h后。

✧ 后续每5000h换油一次。间隔时间不得超过两年。

✧ 油的清洁度极其重要，加油前必须用冲洗油清洗齿轮。注入新油前流体应完全放干。

✧ 不同生产厂商的油在任何情况下都不可以混用，即使可能等级相同。

3.4 操作

- 利用盘车手轮，手工转动驱动滑轮完整的一周，以便润滑油均匀散布在上面。
- 连接钢丝绳前确认曳引机的运行。
- 最初9到10次完整的转动应该半载运行，避免损坏齿轮。
- 一旦曳引机确认完成，以四分之一负载重复测试，接着采用轿厢空载测试。

3.5 制动器调整

所有制动器（如图6-2所示）均由工厂完成设置。除非必须，建议不要更改厂家设置。但是由于任何原因需要调制制动器时，遵守以下流程：

• 打开制动靴时行程尽可能小。

• 利用制动杆打开制动靴。

• 拧紧或松开调整螺栓，直到制动靴和制动鼓之间的间隙为0.1 ~ 0.2mm，利用厚度规进行确认。

• 制动距离取决于弹簧的校准，偶尔需要调整。

• 确认制动靴在正常运转时可以同时打开。

• 定期检查制动靴的磨损程度。

• 如果制动靴磨损显著，根据上述说明调整制动靴。

• 当制动靴材料磨损至2mm或更薄时，必须更换制动靴。确保制动靴或制动鼓上面没有油迹。

■ 4. 维护

定期对机器完成下述普通检查。检查频率取决于机器运转周期。运转周期最高每天2h的每六个月检查一次，运转周期超过每天2h的每三个月检查一次。

4.1 轴向轴承间隙确认

推力轴承的间隙可以通过目测确定，观察换向时制动鼓相对于制动靴的轴向运动量。当这样的间隙变得明显时联系我们的客服部，以确保更换的可能性。

4.2 换油及油位确认

参考3.3润滑

4.3 制动靴磨损确认

参考3.5制动器调整

4.4 油封确认

曳引机具有静态密封（无摩擦）和动态密封（有摩擦）。定期检查曳引机是否漏油，如果需要更换磨损的密封圈，请联系我们的客服部。

4.5 曳引轮槽磨损确认

如果驱动轮槽磨坏，必须更换滑轮。不得在无明确授权下修复轮槽。

4.6 更换零件

根据机器序列号联系我们的客服部。

■ 5. 电机存储、安装及操作指南

• 如果电动机存储时间很长，启动前用500V兆欧表测量绝缘电阻。如果由于不利天气状况（湿度大）测量值很低的话（小于5MΩ），应对电机进行彻底检查。

• 机器必须由被授权人员根据当地电力规则/法规安装。电机应防止过载及

短路状况。

• 电机缠绕引线必须用平底螺母，平垫和弹簧垫圈固定在电极端子板上。

• 根据铭牌上给定的连接方式（星形或三角形）和端子盒盖内的连接图对电机接线。

• 火线应该用环形吊耳连接在电机端子板上。连接电机电源线时，确认电机缠绕引线是否牢固。

• 连接电缆到电机端子时，请确保电缆保护和夹紧合理，不能施加任何拉力在端子上。电缆的重载或拉力能破坏端子。

• 进行有效接地保护。

• 电源电压必须与铭牌上给定的相同或者在指定误差内。

• 检查空载和满载时电机的电流，额定电压时应该小于铭牌标示的10%，不超过铭牌的标示值。

• 在控制回路中使用热敏电阻，避免电机过热。

■ **重要**

✧ 该曳引机设计及制作用于电梯和卷扬机，除此以外的其他用途都是不合理的。

✧ 曳引机不得在规格书规定之外的情景下使用（载荷、速度、钢丝绳、安装等）。

✧ 需求的测试和检查必须由具备资质的人员完成。

✧ 在底座测量的最大工作温度不得超过70℃。

■ 自我评价

1. B　2. C　3. A　4. B　5. D　6. B　7. A　8. B　9. A　10. B

第7章　工程制图

第1节　工程制图简介

工程制图也称为技术制图，用于解释事物如何发挥功能或者如何制作。排列标准和惯例，线条宽度，字体大小，符号，视图投影，画法几何，尺寸和注释用于产生唯一解释方法的图样。

工程图与普通绘图通过如何解释区分。普通绘图可能有多重目的和意义，但工程制图的意图是精准而清晰的交流全部所需规格，将想法转化成物质形态。

■ **建筑制图**

用于形成建筑物的艺术和设计称为建筑学。为了交流设计的全部方面，要用到详细的图纸。建筑制图（如图7-1所示）用来描述和记录建筑师的设计。

建筑制图或者建筑师的图是属于建筑定义的建筑（建筑项目）图纸。建筑制图由建筑师或其他人用于很多目的：将设计概念演变成清晰的提案，交流想法和概念，使客户接受一个设计的优点，使建筑承包商可以建造，完善作品的记录，记录已经存在的建筑物。

■ 机械制图

机械制图（如图7-2所示）是工程制图的一种，用于完整清晰的定义机械零件的要求。

机械绘图（行为）产生机械图纸（文件）。不仅是绘画，还是一种语言——一种人与人之间交流想法和信息的语言。尤其是，它包含了从设计产品的工程师到加工零件的工人的全部所需信息。

■ 服装制图

服装在身体上飘拂。有几个因素导致该现象的产生，即重力和动力。这是你在绘制服装图时要考虑的两种力。 根据姿势和上述因素，服装如何体现在物体上会因人而异。服装制图见图7-3。

■ 剖视立体图

剖视立体图（如图7-4所示）是一种三维制图，图形、简图或说明、三维模型的表面被选择性的移除，从而可以看到内部特征。

剖视立体图避免了空间层次顺序不明确，提供了背景和前景物体的鲜明对比，提供了对空间层次的良好理解。

切掉部分的形状和位置取决于许多因素，例如物体的尺寸、形状和个人取向等。

剖视图和爆炸图是文艺复兴时期的图形发明。剖视立体图一词在19世纪就已经使用，但在20世纪30年代才开始流行起来。

■ 爆炸图

爆炸图（如图7-5所示）是一种展示物体不同零件相互关系或组装顺序的工程图。它展示的物体组件在距离上轻微分开，或者在三维爆炸图中悬浮在周围空间中。一个物体的展示就好像存在一个源自物体中心的小型受控爆炸，导致物体相对于原始位置等距分离。

爆炸图能介绍预期的机械或其他零件的装配体。在机械系统中，一般来说，距离中心最近的组件或者是其他零件装配起来的主要零件最先安装。该图还能表达零件的拆卸，外侧零件一般最先移走。

■ 专利图

专利申请或专利可能包含很多图，也称为专利图（如图7-6所示），用来阐述发明、一些实施实例（尤其是完成发明的实现或方法）或现有技术。法律可能会要求绘图采用特定形式，要求会根据法规而有所不同。

专利图可能包括很多张图，许多视图，版权声明，安全标记和更正（持久的和永久的）。

第2节　机械制图

机械制图（如图7–7所示）是一种根据工程方法产生的工程制图，用于完整而清晰的定义工程项目的要求。工程图一般依据标准的排列惯例，术语，备注解释，外形（字体和线型），大小等。

机械制图能准确而明白地捕捉产品或组件的所有几何特征。机械制图的最终目的是传递能让一个制作者生产零部件的全部所需信息。

"根据标示的信息和规定的尺寸制作此零件，任何未经许可的偏差或错误是你们的责任。"这表达工程图纸完整并合理标注尺寸的重要性。粗心的标注将增加产品成本或使产品彻底报废。

几乎所有的机械制图交流的不仅是图形（形状和位置），还包括尺寸和公差。图样传递下述关键信息：

- 几何形状——物体的形状，以视图为代表，表达物体从不同角度观察时的样子，如前视、俯视、侧视等。
- 尺寸——利用公认单位获得的物体大小。
- 公差——每个尺寸所允许的变动量。
- 材质——代表了物体是什么做的。
- 表面处理——规定物体的表面质量、功能及外观。

■ 视图类型与数量

大多数情况下，单一视图不足以展示所有必要的特征，因此采用数个视图（如图7–8所示）。有六个基本视图，但并不是所有的视图都必须使用。一般来讲，只需要能清晰简洁地传递所需信息的视图。

一般认为，前视图、俯视图和右侧视图是默认情况下的一组核心视图。决定哪个面建立前视图、后视图、上视图和下视图取决于所用的投影方法。根据特定的设计需求，任何视图的组合都可能用到。

除了六个基本视图（前视、后视、俯视、仰视、右视和左视）外，还可能包含一些辅助视图和断面图用于零件定义及其交流。视图线或剖面线（标有A—A、B—B等符号的带箭头直线）定义截面或视图的方向和位置。有时利用小的标记告诉读者在图样的哪个区域能找到该视图或剖面。

剖视图是源对象沿着给定剖切平面的截面的投影视图（辅助或正交）。这些视图跟采用常规投影或虚线相比，能更清晰地揭示内部特征。在装配图中，实心零件（例如螺母、螺栓、垫片）一般不剖。

■ 正交投影

正交投影显示对象跟从前、右、左、上、下或后看上去一样，相互位置关系一般遵循第一视角投影或第三视角投影的原则。投影（也称为投射线）的来源与矢量方向不同，如图7–9所示。第一视角投影是国际标准，主要用于欧洲。第三视角投影主要用于美国和加拿大。

在第一视角投影中，投射光线类似于从观察者的眼球发出，投射到三维物体上在其后方的平面上产生二维图像。三维物体被投射到二维图纸空间，犹如在看物体的射线照片：上视图在前视图下方，右视图在前视图的左侧。

直到19世纪晚期，第一视角投影是北美的标准，跟欧洲一样。大约在19世纪90年代，第三视角的文化传统传遍北美工程和制造领域，达到逐渐广泛接受惯例的程度，在20世纪50年代成为美国标准协会的标准。大约在第一次世界大战时期，英国习惯于混用两种投影方式。

■ 基本图纸尺寸

公制图纸尺寸与国际标准图纸尺寸一致（如图7-10所示）。在20世纪后半段，随着复印变得廉价，图纸尺寸发展得更细化。工程图纸在尺寸上易于加倍（或对半分），与更大尺寸（或更小）图纸放在一起时不浪费空间。

所有的国际标准纸张尺寸具有相同的宽高比，1 ：$\sqrt{2}$，意味着任何尺寸文件可以放大或缩小到其他尺寸并配合精确。考虑到尺寸容易更改，复制或印刷给定的文件到不同尺寸的纸张上就变得很普通，尤其是在一个系列内，比如A3的图可以放大到A2或者缩放到A4。

■ 绘制方式

草图[如图7-11(a)所示]是一种徒手快速完成的绘图，不准备使其成为最终作品。总的来说，草图是一种记录想法以备后用的快速方式。建筑师草图的主要目的是一种提炼不同设想形成作品的方式，尤其是当最终成品非常昂贵或耗时很长。

仪器绘图[如图7-11(b)所示]是将一张纸（或其他材料）放在具有直角和直边的平面上，通常是绘图板。一种称为丁字尺的滑动直尺放在其中一边，允许它沿着桌子边在纸张表面滑动。圆规用于绘制简单的圆弧和圆，三角板用于绘制直线。

现在，手工绘图工作的技术很大程度上已经通过使用计算机辅助设计系统（CAD）[如图7-11(c)所示]提速及实现自动化。有两类计算机辅助系统用于产生工程图纸，即二维绘图和三维绘图。AutoCAD是最早也是现在最流行的二维设计、绘图和建模软件。众所周知的三维绘图软件包括Solidworks, ProE、UG 和Catia。

■ 自我评价

1. D　2. B　3. C　4. C　5. A　6. B　7. B　8. A　9. C

10. geometry、dimensions 、tolerances、material、finish

第8章　产品检测

第1节　公差和质量控制

整个产品的测试是客户接受产品前必须执行的合同约定。测试收集的数据为产品理论设计提供佐证，为设计改进提供解释，以便将来的产品优于现有的产品，以及使设计朝成本更优方向发展。除此之外，它还是一种验证设计的手段，因为并非所有的设计参数可以完全计算出或预估到。

产品测试工程师与设计工程师紧密协作，为测试提供有效数据。他们还必须与制造的各阶段工程师密切协作。产品测试经常可以发现设计缺陷，需要在制造过程中进行重大修改。当公司生产许多样板和小批量生产时尤其如此。所以，制造工程师跟设计工程师一样对产品测试结果有兴趣。

对于复杂的产品，产品测试便成为全部过程控制功能非常重要的一部分。它使公司深信产品表现与客户期望相同，这是一项非常有价值的市场工具，因为它有助于与客户建立良好的声誉。

▪ 极限公差

尺寸控制是非常有必要的，以便确保零组件的装配和互换性。影响间隙和干涉配合的关键尺寸会规定公差。指定公差的方法之一是极限公差。它决定了偏离名义尺寸的多少，可允许的实际测量值，所以一个尺寸可以表述为Φ40.0+/−0.3mm。这意味着改尺寸加工后处于Φ39.7mm和Φ40.3mm之间。

如果变动在名义尺寸的任一侧，公差称为双向公差，如$\Phi 40 \pm 0.03$或$\Phi 40^{+0.4}_{-0.1}$。对于单向公差，一项公差为零，例如$\Phi 40^{0.5}_{0}$或$\Phi 40^{0}_{-0.5}$。实例见图8−1。

很多组织机构会采用通用公差（如图8−2所示）用于图样中没有定义明确的尺寸。对于加工尺寸来说，通用公差可能为+/−0.5mm。所以一个表明为15mm的尺寸可以位于14.5mm和15.5mm之间。其他的通用公差可以用于角度、钻孔和冲孔、铸造、锻造、焊点和圆角。通用公差还有另外一个名字，默认公差。

▪ 公差与配合

配合定义为两个配对物体因尺寸差异造成的关系。公差定义为最大极限尺寸和最小极限尺寸的差值。公差还等于上下偏差的代数差。根据孔或轴实际极限值的不同，配合可以分类如下（如图8−3所示）：

A. 间隙配合，是一种总在配对零件之间提供间隙的配合。在这种情况下，孔的公差带完全在轴的公差带之上。

B. 过度配合，是一种在配对零件之间提供间隙或干涉的配合，取决于成品零件的实际尺寸。

C. 过盈配合，是一种总在配对零件之间提供干涉的配合。这里，孔的公差带完全在轴的公差带之下。

在这种情况下，公差被指定为一个字母（某些情况下两个字母）和一个数字符号，例如20H7或20g6。大写字母用于孔，小写字母用于轴。字母符号表明了公差带相对于零线的位置，零线代表基本尺寸。数字符号代表公差带的值，称为等级或公差质量。位置和公差等级均为基本尺寸的功能。

■ 几何公差

几何公差又称为形位公差（如图8-4所示），一种定义与交流工程公差的系统。它采用工程图样的符号语言和计算机生成的三维实体模型来清晰地描述名义的几何形状和所允许的变量。它告诉制造人员和机械设备，零件的每个面需要何种程度的准确度和精度（如图8-5所示）。

■ 质量控制

从传统意义上讲，质量控制是制造和设计之间的联系纽带。该功能解释为制造和推行质量计划的设计规范，它是综合制造工程方法和操作过程的规划指导。质量控制还负责向管理部门建议容许的制造损耗。传统的质量控制是通过设置不允许超出的预算不良率来控制制造的损耗，从而建立测量和纠错活动的准则。

在过去的一二十年里，质量控制与市场和用户的关系日益紧密，建立文件系统以保证产品质量。这一新的角色成为质量保证的新内容，区别于传统的内部质量控制体系。

质量保证力争通过执行文件和每个制造阶段的特征来确保产品按预期水平完成制造。质量控制与制造过程直接相关，而且质量保证一般还包含在市场调控功能中的用户支持职责。许多工业组织在制造功能下选择建立独立的质量保证子功能，设置了质量控制技术职责，即包含在制造工程组织内的工艺过程控制。

第2节　尺寸测量

■ 卡尺

游标卡尺、转盘卡尺和数显卡尺可以直接给出被测距离的读数，精度很高。它们功能上相同，读数方式不同。这些卡尺包含一对带固定卡爪的刻度尺和另外一对带指示装置且沿着刻度尺滑动的卡爪。

游标卡尺、转盘卡尺和数显卡尺能够利用图中最高处卡爪测量内部尺寸，利用图中下方卡爪测量外部尺寸，很多情况下还可以通过与移动部分连接在一起的探针沿着尺身中部滑动测量深度。该探针细长，能够进入深槽中，而其他测量工具则很困难。

卡爪之间的距离不同，三种类型的读数方式不同。最简单的方式就是直接读出指针在刻度尺的位置。当指针在两个标志中间时，增加游标尺可以使插值更准确，这就是游标卡尺。

游标卡尺（如图8-6所示）可能在刻度尺下部分为米制测量刻度，上部分为英制测量刻度，或者相反。工业常用游标卡尺的精度为0.01mm或10μm，或者千分之一英寸。有些卡尺的大小可以测量72in（1800mm）。游标卡尺的滑动部分通常用一个小的螺钉锁定在某个设定值，判断零件尺寸是否合格。

转盘卡尺（如图8-7所示）没有采用需要练习才能使用的游标机构，它可以在一个简单的转盘上读出毫米或英寸的末尾分数部分。在这种装置中，一个小但精密的齿条来驱动圆形转盘的指针，使得不需要读游标刻度就可以直接读数。指针一般1in、1/10in或1mm转一圈。

现在比较流行的改良是读数显示为单一值的电子数显示屏取代了模拟转盘。一些电子卡尺（如图8-8所示）能够在厘米或毫米与英寸之间切换。全部提供在滑动的任何点位对显示屏归零，与转盘卡尺一样的微分测量方式。数字界面大大减少了测量和记录的时间，提高了记录的可靠度。

普通6in/150mm长数显卡尺是不锈钢做的，标称精度为0.02mm。同样的技术适用于制造8in和12in卡尺。

数字卡尺包含一个电容线性解码器。在滑动块的印刷电路板上蚀刻出条形样式。在卡尺刻度之下的另外一块印刷电路板也含有蚀刻样式的线。两块电路板的组合形成两个可变电容器。这两个电容是错相的。滑动部分移动时，电容值以线性样式和重复模式变化。内置在滑动部分的电路在滑动部分移动时计算条数，并根据电容大小进行线性插值以找到滑动部分的准确位置。

■ 千分尺

使用标有刻度的螺栓而不是滑动条来测量的卡规，称为千分尺或者经常简称为测微计（如图8-9所示）。千分尺一般在规定温度下测量是准确的（一般是20℃或68 ℉，一般称为室温）。

■ 卷尺

卷尺（如图8-10所示）是一种弹性尺，是一种常用的测量工具。它含有一带线性测量刻度的尺身（布、塑料、玻璃纤维或金属片）。它的灵活性允许其测量巨大长度，便于在兜内和工具箱内携带，以及沿曲线和角落测量。测量员使用卷尺能测量100m（300多英尺）的长度。

现在，为裁缝制作的卷尺一般由玻璃纤维做成，不容易撕裂。为木匠和建筑工人设计的测量卷尺多采用坚硬的弯曲金属片，可以在展开时硬而平直，收回时成卷状方便存储。这种类型的卷尺在末端有一个辅助测量的浮动柄舌。柄舌浮动的距离等于自身厚度，所以内侧测量和外侧测量都是准确的。

■ 厚度规

厚度规也称为厚薄规、测隙规或塞尺（如图8-11所示），是多片厚度精准的钢片或塑料片，一般从0.02 ~ 1mm。

典型的厚度规每套中有独立的20 ~ 40片，通过螺栓连接在一起，这个螺栓穿过每片末端的孔。每个量器都标有厚度，可以包含英尺和毫米。当不使用时，全部的扇片折叠进手柄内以免受破坏。

■ 高度计

高度计（如图8-12所示）是一种决定物体高度或在工件上反复标记的测量设备。前一类型的高度计可以用于医生诊断确定人的身高。后一类型的测量工具用于金属加工业或测量学以确定或测量高度距离，指针很锋利可当作画线器或帮助区分工件。

■ 三坐标测量仪

坐标测量系统是一种测量一个物体物理形状特性的设备。这种设备可以由操作员手动也可以由计算机控制。测量通过连接在机器上面的探针实现，探针可以是机械式的、光学式的、激光式的或其他。典型的桥式三坐标测量仪由三个轴构成，即X、Y和Z。

第3节 电气测试

万用表也被称为伏欧表（如图8-13所示），一种集几种测量功能于一身的电子测量仪器。

通过操控切换开关，可以快捷而方便地将万用表设置为电压表、安培表和欧姆表，每种类型的仪表都有几种档位选择（称为量程），还可选择交流和直流。有些万用表还有一些其他功能，如测量晶体管、电容和频率范围。万用表是一种有用的手持装置，用于工业范围和家庭的电气问题查找。

万用表有两种类型，数字型和模拟型。模拟万用表用指针显示，数字万用表用液晶屏显示。

■ 模拟万用表

模拟万用表的分辨率受到指针宽度、指针抖动、刻度印刷的精确度、调零、量程数和机械式显示类型读数对不水平造成的错误等因素的影响。

尤其是电阻测量回路造成电阻的测量精度低。廉价的指针万用表可能只有一个电阻刻度，严重限制了测量精度的范围。典型的模拟万用表具有一个面板欧姆调零装置，以补偿万用表电池的电压变化。

■ 数字万用表

数字万用表是我们家庭和工作中最有用和最有帮助的工具之一。拥有一个好的型号并学会如何正确的使用是非常重要的。今天，当代的数字万用表都设计得坚固易用。一个好的万用表应具有坚固的塑料外壳和方便使用的大旋钮。

数字万用表的上部是一个数字读数显示器，这个是在购买之前必须仔细检查的。确保屏幕足够大以便于读数，而且在日光下应该可以看到读数，因为在户外使用时可能遇到阳光直射。确保功能开关大且操作简单，大多数功能开关有八个位置。

在放回工具箱之前关掉设备的电源。这样做的目的是节约电池，因为下次还会用到它。

■ 万用表的使用

万用表使用不当容易损坏，请采取以下预防措施：

1. 根据预期的读数范围选择大一级的量程。

2. 连接仪表，确认万用表的表笔以正确的方式形成回路。如果接反了，对于数字表可能是安全的，对于模拟表可能损坏。

3. 如果读数超出刻度范围，立即断开，选择更高一级量程。

4. 在调节量程前，保持万用表处于断开状态。

5. 在测量前检查量程设置。

6. 除非实际要测电流，绝对不能把万用表设置在电流挡。因为万用表的电阻较小，所以在电流挡时最容易损坏。

当进行电路测量时，常常需要知道某一点的电压。例如图8-14中555定时器芯片2脚的电压。这似乎有点让人感到迷惑，应该把万用表的表笔放在哪里呢?

将黑表笔接在0V处，通常是电池或电源的负端。然后将红表笔放在需要测量电压的点上。可以将黑表笔固定在0V位置，同时用红表笔依次在各个点测量电压。

数字万用表的另一个有用的功能是欧姆表。欧姆表是测量电阻的仪表，如果电路没有电阻，则欧姆表读数为0；如果电路断开，则欧姆表读数为无穷大。

■ 自我评价

1. A　2. A　3. B　4. C　5. B　6. D　7. B　8. A　9. B　10. D

11. internal dimensions、external dimensions、depth

第9章　电气系统简介

第1节　电路

■ 什么是电路

图片展示的是电路的基本类型，电路可以采用如图9-1所示的图进行描述。它包括电源，消耗能量的负载及连接电源和负载的导体。

只要如图显示，从电源到负载和回来的连接没有断开，电荷会从电源负极经过负载回到电源正极。箭头显示电路流经回路的方向。因为电荷经常沿着回路同一方向运动，这就是所知的直流电。

电源可以是任何形式的电能来源。实际上可能有三类：电池，发电机或某些形式的电源。负载可以是电力驱动的任何装置和回路。它消耗来自电路的电能，转化成产品——热量或光。负载可以简单到一盏灯泡，或者复杂到一台现代高速计算机。

■ 电路种类

存在三种类型的电路：串联电路，并联电路和串并联电路。串联电路是最简单的，因为电流只可能流经一条回路；如果电路断开，任何负载装置都不工作。并联电路的不同之处在于它包含电流流经的路径多于一条，所以其中一条路径断开，其他路径继续工作。串并联电路是上述两者的结合，一些负载在串联电路，其他的在并联电路。

模拟电路的电流和电压可能一直随时间根据所代表的信息而变化。模拟电路由两个基本模块构成，串联和并联电路。在串联电路中，流经一系列元器件的电流相同。圣诞灯绳就是串联电路的好例子，如果一个熄灭，其他的也会熄灭。在并联电路中，所有元器件的电压相同，电流根据元器件的电阻进行分割。

在数字回路中，电子信号以离散值标示逻辑和数字值。这些数值代表了被传递的信息。在大多数情况下，采用二进制码：一个电压，通常是正值，代表二进制的1，另外一个电压，经常是接近零电势，代表二进制的0。

混合电路包含模拟和数字电路。实例包括比较器，定时器，模拟数字转换器和数字模拟转换器。多数的现代收音机和通信电路采用混合信号电路。例如，一个接收器，利用模拟电路放大和频率转换信号以达到合适状态再转换成数字值，然后进一步信号处理可以在数字领域进行。

■ 电压和电流

电源提供的电力具有两个基本特性，称为电压和电流。

电压被称为是电势能，测量单位是V，是两点的电势差或两点单位电荷的电势能差。电气压力促使自由电荷在电路中移动，称为电动势。

电压表可以用于测量系统中两点的电压（或势能差），其中一点采用常见的参考电势，如系统地。电压的产生可以是静态电场，电流流经磁场，随时间变化的磁场，或者上述三者的综合。

电流是电荷沿着媒介的流动。电荷一般由电线的导体中移动的电子携带。它也可以由电解液的离子，等离子中的离子携带。

测量电荷流动比率的国际单位是A，是电荷以1C/s的速度流经某表面。也就是电流的测量单位是A。

■ 欧姆定律

许多定律适用于电路，但欧姆定律应该是最广为人知的。欧姆定律指出，电路的电流与电压成正比，与电阻成反比。所以，如果电压增加则电流增加；电阻增加则电流降低。

为了理解欧姆定律，理解电流、电压和电阻的概念就非常重要：电流是电荷的流动，电压是驱动电流同一方向运动的动力，电阻标示物体阻碍电流通过。

欧姆定律公式为$E = IR$，其中E代表电压，V；I代表电流，A；R代表电阻，Ω。该公式可以用于电路分析电压、电流和电阻。

第2节　电器元件

电器元件或电子元件一般具有两个或多个引脚的离散形式（如图9-2所示）。它们一般通过焊接到印刷电路板连接到一起，以产生电流用于特定目的，例如放大器、收音机听筒或振荡器。电器元件可以单独包装或者类似薄膜装置的集中包装。

■ 二极管

大多数二极管（如图9-3所示）在外观上与电阻类似，在一端的漆线表示方向或流向（白色一侧是负极）。

二极管基本上是电流的单向阀。只能允许电流一个方向流动（从正极到负极），而另一个方向不能流动。如果二极管的负极在电路的负极端，电流流动。如果二极管的负极在电路正极一侧，没有电流流动。

■ 发光二极管

发光二极管（如图9-4所示）是一种电子光源。发光二极管的基础是半导体二极管。当二极管正向偏压（打开），电荷与空穴复合，能量以光的形式释放出。这种效应称为电致发光，光的颜色由半导体的能量差决定。

发光二极管具备传统光源所不具有的许多优点，包括低能耗，寿命长，鲁棒性改善，尺寸小。

■ 晶体管/三极管

晶体管（如图9-5所示）是现代电子学的关键元件。许多人认为它是20世纪最伟大的发明之一。有NPN和PNP两种类型，电路符号不同。

晶体管的重要用途来源于其能力，使用一对端子之间的小信号控制另外一对端子之间的大信号。这种特性称为增益。晶体管可以根据输入信号成比例的控制其输出信号。也就是说，它可以作为一个放大器。晶体管能作为一个电子控制开关打开或关掉电路中电流，电流数量由其他电路元件决定。

■ 晶闸管

晶闸管（如图9-6所示）是一个固态的半导体装置，具有四层交替的N型和P型材料，起双稳态开关的作用。

■ 电阻

电阻器（如图9-7所示）抗拒电流的流经。电压不变时，电阻值越大（测量单位为欧姆），电流越小。

■ 电容

电容器（如图9-8所示）是一种在电子领域存储能量的被动电子元件。实际电容器的外形变化很多，但都具有至少两个被绝缘体隔开的传导装置。电容的作用类似于电池，但充放电效率更高（电池可以储存更多的电量）。

■ 继电器

继电器是一种电子操作开关。流经继电器线圈的电流产生磁场，吸起杠杆而改变触点。线圈电流可以通断，所以继电器有两个开关位置，是双投（转换）开关。

第一批继电器用于长途电报回路，重复来自一个回路的信号并再次传递给另一个回路。继电器广泛用于电话交换机和早期的电脑用于执行逻辑运算。

■ 单片机

单片机（如图9-9所示）是集成在一个芯片上的完整计算机系统。尽管所有特征都在一个小的芯片中，但其具有大部分计算机元件，如CPU、存储器、内部和外部总线系统。它同时集成了通信界面，定时器，时钟和周边设备。现在最强大的单片机甚至可以集成声音，图像，网络，复杂的输入和输出系统在一个芯片中。

如果你想在因特网上得到一些关于单片机的英文信息，你认为关键字应该是什么？ single chip microprocessor, SCP, microcontroller unit, microcontroller 还是MCU？因为microcontroller unit比single chip microprocessor更常用，而且缩写可能有多个意思，所以最好的关键字是microcontroller unit和microcontroller。如你所知，合理的关键字可以缩小范围，利于找到你想要的信息。

■ 可编程控制器

可编程控制器是一种用于机电自动化的数字电脑，例如工厂组装线的机器控制。与通用计算机不同，可编程控制器设计用于多输入和多输出，适用温度范围大，对电气噪声有免疫力，可抵抗振动和冲击。

PLC（如图9-10所示）的发明是为了响应美国汽车制造业的需求。可编程逻辑控制器最初被汽车行业采用，当生产型号变更时，软件修改代替了硬接线控制面板的重新接线。

在PLC之前，汽车制造的控制，排序和安全联锁逻辑通过成百上千的继电器、凸轮定时器和专用闭环控制器完成。每年型号变更的设备更新过程非常耗时且昂贵，因为电工需要对每个继电器单独重新接线。

1968年，通用汽车的自动传动部门发布了硬接线继电器系统电子取代的方案征询。最初的PLC就是获奖方案，被指定为084是因为这是贝德福德协会的第84个项目。贝德福德协会成立一间新公司从事该新产品的开发、生产、销售和服务：Modicon，代表模块数字控制器。

迪克·莫雷是从事该项目的人员之一，被认为是PLC之父。Modicon商标在1977年被卖给高德电子，后来被德国公司AEG接手，接下来是法国施耐德电气，即现在的所有者。

■ 自我评价

1. B　2. A　3. B　4. B　5. C　6. C　7. B　8. A　9. D　10. B

第10章　电梯安装与维保

第1节　电梯安装

安装一部新电梯时，安装人员应首先研究图纸以确定所需的设备。准备完并确认后，才能开始安装。

电梯安装人员沿着井道壁放置导线管。一旦导线管就位，机械师会将塑料外壳的电线穿过导线管。接下来就可以在每层和机房内安装所需电气元件及相关装置。

安装工人通过螺栓或焊接将导轨与在井道壁上的导轨支架固定在一起。电梯轿厢一般在井道上方安装，导靴和安全钳会固定在轿架上。安装人员还需要在每层的出入口安装厅门和门框。

技术最精湛的安装人员称为调试工，专门负责安装后设备的精细调整。调试工要确保电梯根据要求运行，并且在指定时间内准确的停在每一层。调试工需要充分了解电子学，电力和计算机，以确保新装电梯正常运行。

■ 总体确认

在开始安装前确认以下事项：

① 确认井道是否方正与垂直；

② 比对井道实际尺寸与总体设计图提供的尺寸，如宽度和深度；

③ 确认全部行程（从楼层面到楼层面的距离）；

④ 确认底坑深度；

⑤ 确认顶层空间（从楼层上表面量到天花板下方）；

⑥ 确认井道中导轨支架安装的固定位置。

■ 物料交付

根据发运清单核对所有运货部件，以确认是否存在少料。包装箱的任何损伤需要报告给运输公司和供货商，尽快给他们发一份备忘录，列出损坏的设备。

■ 安装进度

安装项目的时间随项目不同而变化，取决于下述因素。

① 需要安装的楼层数；

② 机器方位；

③ 选择自动门操作还是手动门操作；

④ 场地条件和安装队的经验；

⑤ 入口由电梯承包商还是其他人安装。

在安装队开工前，监管员视察现场以确认尺寸正确和所有服务设施齐全是非常重要的。如果脚手架和电力不具备，工作不应开展。通常来说，第一天就需要脚手架，在第一天末或第二天初需要电力带动机械。

第2节　装梯安全

■ 总体安全注意事项

尽管安全是每个人的责任，但主管技术员具体负责协调工地各处位置的安全。保持工作区域无废料、垃圾和其他可能引起绊倒危险的废弃物。清洁时戴上皮手套并使用簸箕和扫帚，以避免被藏在垃圾中的尖锐物体割破或刺穿。

将沾满油污的破布放入许可的容器并定期从工作场地移走，以降低风险。将木板上的钉子移走或者扳平，注意绊倒风险。以不同的方式堆叠用过的材料，以便于移走时识别。损坏的钢丝绳不得使用，并从工作现场移走。

在工作现场，安全最重要，所以，不允许使用收音机和磁带播放器。很自然地，禁止喧闹，恶作剧，打闹和持有武器。

当雇员的能力和警戒性受到因以下因素削弱时不应工作：

① 疲劳；

② 含酒精的饮料；

③ 非法或处方药；

④ 其他会降低注意力或抑制能力的因素。

在相邻电梯运行的多井道内对电梯进行安装或改造时，施工的井道部分需要隔离。

在有公众出现的地方，应该有至少8ft高或到天花板的坚固栅栏封闭工作区域。将损坏或维护不合理的护栏及栅栏立即汇报给监察员和总承包商。关于护栏和栅栏要求的其他信息请参考安全手册。

安全操作是工作中最重要的部分。安全是每个人的责任，应及时提醒有不安全行为的人。同样，以感激的心情接受别人的提醒。持续翻阅电梯行业员工安全手册，公司的安全政策与程序，以及工作现场要求是安全责任的重要一部分。

■ 电路操作

在做需要与导体接触的任何工作，例如更换电气元件时，总是断开电路，接着用仪表确认电压。还要意识到在多轿厢安装时，即使一个轿厢的主线断开，控制器还可能有电压。

当站在水中或金属表面时，绝不可以操作电路。记住，电梯主线断开时并没有对底坑照明，排水泵或轿顶轿厢照明断电。

不要在口袋中携带工具。这样做增加了工具接触带电回路的风险，应挂在移动设备上或者递给下方的员工。

当接近或操作带电回路时，你应该：

① 站在干木板或者橡胶垫上；

② 从不站在金属或潮湿表面；

③ 只使用绝缘工具和工作灯；

④ 摘掉所有的首饰和钥匙环；

⑤ 佩戴非金属边框眼镜；

⑥ 安装保险丝时利用拔丝钳并确认断电；

⑦ 不要用跳线短接保险丝；

⑧ 当工作结束后更换外壳；

⑨ 在控制器检修开关旁放一块标签，内容如下：出入口及安全回路不得短接。

■ 劳保防护

应当按照图纸建造隔离装置。公司提供的安全帽应该与头部合理匹配并固定。

从事以下工作时应当佩戴合适的眼睛防护工具：

① 焊接；

② 切割；

③ 浇注巴氏合金；

④ 使用化学试品或溶剂；

⑤ 在满是灰尘区域工作。

有潜在危险时，如处理重物或原材料，应戴手套。处理钢丝绳时要一直戴手套。靠近移动或旋转机械装置时绝不可以戴手套。

■ 梯子

只能使用公司下发的带有安全脚的梯子。金属梯子导电，不得使用。梯子不得油漆，因为油漆会掩盖裂纹和其他缺陷。损坏或有裂缝的梯子应贴上“不得使用”标签并从工作现场移走。

台阶梯子应该全部打开并合理使用。如果使用的梯子可能被该区域的其他人碰到，应该有另外一个人一直在梯子下方（防止梯子被撞倒）。不要在这种情形下让梯子无人看管。如果梯子放在过道或走廊，应当设置障碍或用绳子围起来。

在梯子上工作时，避免伸出超过一个胳膊的长度。探身出去能导致不平衡，置自己于危险位置，有坠落风险。永远不要踩在梯子的顶部台阶上。

■ 仿形轿厢

有各种各样的仿形轿厢和类似的装置，用于在井道内对接导轨、固定支架、安装入口、走线和其他安装电梯需要的工作。

使用仿形轿厢的一般规则如下：

① 每天使用前进行静态安全测试，并确认顶层支撑，所有连接和绳索。

② 记录所有的安全测试，包括轿厢投入使用和移到不同井道的测试。

③ 只用来运载电梯工人，不得提升重物。

④ 要意识到载重有限，不要超过界限。

⑤ 在仿形轿厢上工作的人员应该穿好固定在独立生命线上的全身安全带和吸振绳索。

⑥ 需要顶层保护时，推荐使用开孔1/4in大小的安全网。

■ 坠落防护

在一些情况下，在梯子上、仿形轿厢、其他施工和服务的区域，需要有人体坠落捕捉系统。在高于地平面6ft（1.8m）或开口大于10in（254mm）的地方工作，工人处于坠落危险时必须采用坠落防护。

在井道内或周边工作时，必须使用护栏系统。这样做不仅是保护自己，也是保护可能在此区域工作的其他人。具有坠落风险的设备应该停止使用并销毁。

以下是使用坠落捕捉系统时必须遵守的规则列表：

① 只能使用公司许可的救生索、吸振绳索和身体安全带。

② 救生索应该免受切割和磨损。

③ 每根垂直的救生索只许一人使用。

④ 禁止直接与曳引绳固定，必须使用合理的绳索固定装置。

⑤ 救生索必须在井道施工前装好，应可以贯穿井道全部长度。

■ 手拉葫芦

每天使用手拉葫芦前，目测检查是否存在缺陷。近距离观察链条和底部的挂钩。底部的挂钩是整个链条的最薄弱部分。底部挂钩损坏和伸展是过载的征兆。测量底部挂钩的伸展量，与安全手册中的最大允许量比对。绝不能更换，加长或拼接链条，或者超过链条的承载能力。维修只能由生产商或授权代理商完成。

■ 气罐

乙炔罐和氧气罐承受高压，处理时必须非常细心。以直立位置存储氧气和乙炔罐，实际使用时离开20ft远，或者通过不小于5ft高的防火墙隔开。

绝不润滑测量仪器和罐体的螺纹。小心注意软管以防止损坏，不要使用已磨损的软管。

■ 焊接预防措施

开始焊接前，清理所有的易燃材料。如果易燃物不能移走，要用耐火材料覆盖。特别注意油和溶剂。木地板必须润湿或用金属板及类似材料盖住。应当注意火花可能溅落的位置，因为可能引起火灾或者溅到其他工人身上。

焊接时穿恰当的衣服。绝不要穿油渍或者易燃衣服。另外一个配有ABC型灭火器的工作人员在焊接时站在旁边留意火情。焊接结束后，需要对该区域和临近区域监控一段时间。

焊接时合理的眼睛防护是非常重要的。应该在电弧焊时使用具有头部保护的焊工头盔。

■ 钢丝绳紧固

推荐在可能情况下采用楔形绳头组合。然而，如果采用锥形座，要遵守下述注意事项：

① 浇注巴氏合金时，佩戴面具并戴好手套。

② 在向座套中注入熔化的巴氏合金时，对座套预热以确保没有水分存在。

③ 只能在通风良好的地方工作，避免吸入烟雾。

④ 处理完巴氏合金后吃东西或吸烟之前需彻底洗手。

采用绳夹连接钢丝绳时，参考安全手册确定所需的绳夹数量和之间的距离。将第一个绳夹放置在距离钢丝绳末端一个夹鞍宽度的地方。将U形螺栓放在终端一侧，鞍部在活动端。

均匀地拧紧螺母，两只螺母交替进行，直到达到推荐的扭矩。推荐扭矩可以在钢丝绳使用手册或绳夹供应商目录中找到。一般来说，对于3/8in的钢丝绳为45lbf，对于1/2in的钢丝绳为65lbf。螺牙必须清洁、干燥，无润滑油时扭矩才能准确。施加预期的载荷以检查连接情况。

夹得过紧将导致钢丝绳变形或受损，削弱连接。反之，夹得过松时钢丝绳可能滑动，损伤钢丝绳，削弱连接。钢丝绳连接使用几天后，检查移动量和损伤，夹紧到推荐扭矩值。

第3节　电梯维保

一旦电梯投入使用，必须定期维护以保持其在安全工作状态。电梯维修人员一般做预防性的维护，例如加油和润滑移动部件，更换磨损的零件，利用仪表测试设备，调整设备优化性能。他们确保设备和机房干净。他们也会解决问题，如做紧急维修。专门负责电梯维护工作的人员一天中大多数时间独立工作，多数情况下维护多台相同的电梯。

维保人员也要处理主要的维修，例如更换电缆、电梯门或机器轴承。这些工作可能需要使用切割机或装配设备，电梯维修工一般不会携带这些工具。维保人员还要做一些更换工作，比如搬运和更换电机、液压泵和控制面板。

- 自我评价

1. B　2. C　3. D　4. B　5. A　6. B　7. A　8. A
9. A　10. B　11. D.A　12. A　13. A　14. A　15. C

第11章　扶梯和自动人行道

第1节　扶梯

扶梯能够在短时间内输送大量人员。所以，它们广泛用于运输系统、地铁和机场。同样的原因，它们还用于学校和办公大楼。由于具有运送人员不费力，困难小以致无困难的能力，扶梯还广泛用于百货商店和购物中心。

额定速度定义为额定负载时沿倾斜角度上升的速度。美国扶梯的一般速度为90 ~ 100ft/min（0.45 ~ 0.51m/s）。A17.1—2000 (6.1.4.1)将扶梯的最大额定速度从125ft/min降到100ft/min。购物中心和百货商店的扶梯运行速度比地铁设施的略低。扶梯必须通过钥匙操作开关启动，禁止任何方式的自行启动。

40in宽、运行速度90ft/min的扶梯上，每个台阶站一个人，理论输送能力为8102人/h，但额定输送能力大约是其一半（备注，台阶前后距离15.75in）。扶梯目的并非是运送残疾人士、婴儿手推车和大件物体，也包括行走能力受影响，如关节炎、进出扶梯有困难的能力不健全人员。所以，在有扶梯使用的地方仍然需要电梯。

■ 扶梯零件及其结构

具有实心护板的高盖板扶梯零件如图11–1所示。停止按钮在扶梯两端一览无遗。同样的组件设计适用于自动人行道，除了梯级改称为踏板。在过去，停止按钮位于靠近启动开关的前面板处，不容易看到的地方，因而很多人并不知道它们在那儿。

玻璃护板的低盖板扶梯零件如图11–2所示。两种扶梯都为A17.1法规要求的警示标志留出了一处位置。A17法规要求的警示标志如上图所示。

从图11–3和图11–4可以看出，扶梯是一种复杂的机械装置，需要受过培训并具有资质的人员进行维保。

■ 安全装置

扶梯每天被数千人使用。全世界负责任的业主都尽自己所能，通过升级现存的垂直运输设备以满足当地和国家的安全法规来确保人们的安全。他们通过使用安全产品使行人免于潜在的灾难和伤害。

现代扶梯设计的尽可能安全，具有众多发现不安全状态的安全装置。这些装置如图11–5所示。

1) 在维修或检查移走盖板时，盖板开关阻止扶梯运行。也就是说，盖板移走时，盖板开关被触发，扶梯关机。

2) 机器的停止开关用于保证维护人员的安全。

3) 裙板障碍开关位于每端的两侧。当台阶靠近梳齿板时，有物体卡入台阶和裙板之间，该开关将在卡入物体接触梳齿板时制停扶梯。

4) 如果梯级链断裂，扶梯将会停止，这就是梯级链断裂装置的作用。在大多数情况下，停止意味着断电。台阶可能在重力作用下继续运动，因为没有类似电梯中安全的装置。

5) 如果扶梯速度超过预设值，速度监控器将制停扶梯。交流感应电机直接与驱动主机相连时，不需要速度监控器。

6) 在运转过程中运动换向时，逆转制停装置将制停扶梯。

7) 如果扶梯台阶靠近下曲线时向上移动，梯级防跳装置将制停扶梯。

8) 如果电机不通过连续轴、机械耦合或者齿轮装置与减速齿轮相连，耦合方式失效时，该装置制停扶梯。例如，该装置需要D形带或链式联轴节。

9) 如果驱动主机通过链传动连接到主传动，该装置将检测断链并制停扶梯。

10) 梯级水平装置可以发现上升端任一侧1/8in的下降位移。该装置会在该位移进入梳齿板区域之前制停扶梯。

11) 梳齿板–梯级冲击装置受到水平或竖直方向的力将制停扶梯。允许水平力一侧不超过400lb/ft或中心不超过800lb/ft，垂直力不大于150lb/ft。

12) 如果任一条扶手带速度变化过大，超过2s，扶手带速度监控装置将制停扶梯。扶手带停止装置应设在速度监控装置之前。

13) 梯级缺少装置会在缺少梯级的位置出现在梳齿板之前制停扶梯。你可以发现，大多数安全装置位于上下两端，而中间位置空缺。

14) 主线断开将断开扶梯所有电力。

15) 检修开关允许维修时连续按压运动。

16) 在两端梯级下方的梯级边界灯，或梯级与梳齿板间的可视对比物提醒行人注意悬空的边界。

17) 如果扶梯配有卷闸门以关闭扶梯，闸门上开关在闸门开始关闭时将制停扶梯。

18) 扶手带夹手防护：需要在扶手带进入中心柱位置提供防护，避免手与扶手带一同进入护板。该装置可以是弹性靴、刷子或陷阱门。从1992年开始，A17.1法规要求有扶手带入口装置，在物体靠近扶手带与防护装置之间或者卡在该区域时，制停扶梯。

19) 防滑装置：防滑装置用于防止行李、箱子和人员的滑动。

20) 盖板防护：阻止进入外盖板的障碍，防止人员进入及跌落。

21) 安全区：A17.1法规要求入口和出口保持无障碍的安全区域。

22) 装在裙板或内盖板上面的安全刷用于提醒乘客过于靠近扶梯梯级和裙板之间的间隙。

扶梯基本设计结构和尺寸基于安全第一而建立。设计方面的改进和A17.1的要求不断改善。最近一项变化涉及降低陷入梯级两侧和裙板的方法；建立了间隙和裙板摩擦系数的性能指标；而且承认了使用反射器以防止与裙板接触。法规还确认了随梯级运动的动态裙板，以降低卡住的可能性。

第2节　自动人行道

■ 自动人行道类型

自动人行道（如图11-6所示）具有很多与扶梯相同的安全设计，应等同对待。自动人行道斜度最大12°（扶梯一般是30° +/-1°）。主要有两种类型的自动人行道在生产：托盘式和连续带式。

托盘式扶梯包含多块用链条连接到一起的独立托盘，与扶梯非常类似。托盘可以是铸铝或者不锈钢的，宽度在32ft（800mm）和56ft（1200mm）之间。速度一般为100ft/min（0.5m/s），由交流感应电机驱动。连续带式：用一种厚橡胶皮带在两侧通过皮带下方的滚子支撑，称为侧边支撑。

两种类型的自动人行道都具有沟槽表面，在扶梯末端同梳齿板相啮合。而且，几乎所有的自动人行道都具有与扶梯类似的移动扶手带。

■ 倾斜式自动人行道

倾斜自动人行道用于机场和超市运送人员到另一楼层，具有电梯的便利（也就是人们可以随身携带行李车或购物车、童车）和扶梯的输送能力。

手推车或者具备刹车装置，刹车装置在手推车把手释放时自动启用，牢固的磁力让轮子附着在底板上，或者采用特制的轮子使手推车处于斜坡槽中，以便轮子骑在凸起处不至于滑走。

■ 高速自动人行道

在20世纪70年代邓洛普开发了该高速系统。相对于现存的系统，高速的最大优点在于出入区域宽而且移动慢（4名乘客可同时进入，最大允许乘客量达每小时10000人），但是运送区域较窄而且移动速度快。

系统入口类似一个非常宽的扶梯，具有平行四边形形状的宽大金属踏板。经过短距离后，踏板被加速到一侧，滑到另外一个上，形成逐渐变窄但快速运行的轨道，与入口区域几乎成直角运行。乘客经过一个抛物线路径被加速到最大设计速度15km/h（9英里/小时）。

这种体验对乘客来说并不熟悉，他们需要理解如何使用该系统是安全的。为现在的系统开发一种移动扶手带是一种挑战，已被巴泰尔团队所解决。该高速通道打算用于短途独立系统，为传统与之并行的人行道出入口提供加速和减速的方法。该系统开发于1975年，但从未投入商业生产。

2002年，第一条成功的高速通道安装在巴黎的蒙帕纳斯地铁站（如图11-7所示）。起初运转速度为12km/h（7英里/小时），由于人们易失去平衡，速度降低到9km/h（6英里/小时）。据估计，经常乘车的人如果每天使用该通道两次，每年会节省10h或每周15min。

使用高速通道跟其他自动人行道相同，除了为安全而设的进入及离开时的特殊步骤。刚引进该通道时，穿黄色夹克的工作人员决定谁能和谁不能使用。行人必须至少有一只手空闲，以便握住扶手带，而那些携带箱包、购物之类的人，或体弱人员必须使用旁边的普通通道。

■ 自我评价

1. A　2. D　3. C　4. A　5. D　6. A　7. B　8. C

Appendix II

Lift Common Vocabulary
电梯常用词汇

☞ A

AC drive 交流拖动
AC feedback control 交流反馈控制
AC motor 交流电机
AC servo 交流伺服
AC servo motor 交流伺服电机
AC single speed 交流单速
AC two speed motor 交流双速电动机
access forbidden 禁止入内
access lift 通道电梯
accessible 允许进入
accessible space 允许进入的场地
accident 事故
* AC-GL machine 交流无齿曳引机
ACVF drive 交流调频拖动
ACVF system 交流调频系统
* ACVF(AC varible frequency) 交流变频调速
ACVV dirve 交流调压拖动
ACVV system 交流调压系统
* ACVV(AC varible voltage) 交流调压
adjacent car 相邻轿厢
adjacent entrance 相邻出入口
adjusment 调试
** adjustor 调试员
after sales service 售后服务
* alarm button 警铃按钮
angle guide 角铁导轨
angle of lead 导程角
AC variale speed 交流调速
AC (alternating current) 交流电
acceptance of lift 电梯验收
acceptance period 验收阶段
acceptance test 验收试验
acceptance certificate 验收证书
* access door 检修门
angle of traction 曳引机包角
angle of wrap 包角
* ANSI(America Nationa Standard In-stirute) 美国国家标准协会
anti-rebound of compensation rope device 补偿绳防跳装置
* arrival alarm 到站钟
armature winding 电枢绕组
assemble 装配
attendant control 有司机控制
automatic door 自动门
automatic homing 自动回基层
automatic landing 自动停站
automatic leveling 自动平层
automatic telescopic center opening sliding door 自动中分式折叠滑动门
automatic telescopic sliding door 自动折叠式滑动门
* automobile(car) lift 汽车电梯
* auxiliary brake 附加制动器

☞ B

back plunger type 后部柱塞式(液压梯)
back type governor 轴流式限速器
* bearing beam 承重梁
* balustrades 扶手装置
* bed lift 病床电梯
bed plate 底座
belt type moving walk 带式自动人行道
* bi-parting door 垂直中分门
block of flats 住宅楼区
boading floor 登梯楼层
** brake 制动器
brake coutact 制动器联轴器
brake dish 制动盘
* brake drum 制动轮
brake energy 制动力
brake spring 制动弹簧
brake tension 制动器张紧力
brake torque 制动器力矩
** brake wrench 制动器扳手
braking 制动
braking distance 制动距离

broken step chain contact 断绳触点
broken drive chain contact 主驱动链保护装置
* broken drive-chain safety device 驱动链条保护装置
broken rope contact 驱动链条列断安全装置
broken step chain device 断绳开关

☞ C

cab interior 轿厢内部
cage 轿厢
calling landing 呼梯
* calling board 召唤盒
call-out response time 维修召唤应答时间
captive roller safety gear 滚柱式安全钳
* car 轿厢
car annunciator 轿底报警器
car axis 轿厢中心线
car back wall 轿厢后壁
car botom clearance 轿底安全高度
car bottom overtravel 轿厢底部越层
car bottom runby 底部越程
car buffer 轿厢缓冲器
car button 轿厢按钮
* car cab 轿厢体
* car ceiling 轿厢装饰顶

car call button 轿厢指令按钮
car depth 轿厢深度
car dispatch 轿厢调度
car door 轿门
car door catch 轿门栓
** car door lock 轿门门锁

car switch 轿厢开关
** car switch automatic floor stop operation 轿厢开关自动停站装置
car switch control 轿厢开关控制
car switch opration 手柄开关操纵
car top 轿厢顶部

braking force 制动力
** buffer 缓冲器
buffer base 缓冲器底座
* buffer plate 缓冲器板
buffer plunger 缓冲器柱塞
* buffer return spring 缓冲器复位弹簧
buffer stand 缓冲器台
* buider's work drawing 土建图纸
bumper 弹簧式缓冲器

* car door rail/track 轿门导轨
car door sill 轿门地坎
* car emergency exit 轿厢安全门、应急门
* car emergency opening 轿厢安全窗
car entrance 轿厢入口
car extetior 轿厢外装修
car fan 轿厢风扇
** car frame 轿架
car gate 轿厢门
car guide 轿厢导轨
car guide axis 轿厢导轨中心线
car height 轿厢高度
car illumination 轿厢照明
car leveling device 轿厢平层装置
car light 轿厢照明装置
car lighting 轿厢照明
car overload 轿厢超载
** car operation panel（COP）轿厢操纵面板
car panel attachment 轿厢壁板附件
** car platform 轿底
* car roof 轿顶
car door guide shoe 轿门导靴
center opening vertical sliding door 中分式垂直滑动门
* chain 链条
** chain block/hoist 手拉葫芦
clear hoistway 井道净尺寸
center plunger hydraulic lift 柱塞直顶式液压电梯

car top clearance 轿顶安全高度
car top emergency exit 轿顶安全出口
* car top guard 轿顶护栏
car top inspection device 轿顶检修装置
car top inspection station 轿顶检修盒(站)
* car top light 轿顶照明装置
car top protection balustrade 轿顶防护栏杆
* car top protection railing 轿顶防护栏杆
car ventilation 轿厢通风
** car wall/enclosure 轿壁
car weight 轿厢重量
car width 轿厢宽度
center opening door 中分门
center opening two speed door 中分四扇门
car sheave 轿厢绳轮
car side wall 轿厢侧壁
car sill 轿厢地坎
center-opening folding door 中分式折叠门
clearance between skirt panels 裙板间隙
compensating chain device 补偿链装置
* compensating device 补偿装置
compensating rope device 补偿绳装置
** comb plate 梳齿板
comb safety device 梳齿板安全装置
** concrete 混凝土
** control cabinet 控制柜
control cable 控制电缆
** counterweight 对重
* coupling 联轴器
cross-section 横截面

☞ D

** dumbwaiter 杂物梯，餐梯
** door operator 门机
door interlock 门锁装置
* door protection device 门保护装置
door closing protection 关门保护
* deflector sheave 导向轮
* direction reversal device 逆转保护装置
* diversion sheave 反绳轮
** deck 盖板
* driving machine 驱动主机

☞ E

** elevator 电梯
* emergency unlocking device 紧急开锁装置
** escalator 扶梯

☞ F

* freght lift 载货电梯
folding door 折叠式门，栅栏门
* foundation bolt 地脚螺栓
** fish plate 导轨连接板
** floor plate 楼层板，踏板
full load 满载
full load current 满载电流
friction angle 摩擦角
* firman switch 消防开关
** frequency inverter 变频器

☞ G

* geared machine 齿轮曳引机
* gearless machine 无齿曳引机
gear reducer 齿轮减速装置
* goods lift 货梯
* groove profile 槽型
** guide rail 导轨
* (guide) rail bracket 导轨架
** guide shoe 导靴
guide shoe bush 导靴衬
governor 限速器

☞ H

** handrail 扶手带
handrail entry guard 扶手带入口保护装置
** handwheel 盘车手轮
* hall door 厅门
** hall button 召唤盒
* hilti bolt 膨胀螺栓
hydraulic buffter 液压缓冲器
hoist machine 提升机，曳引机
hoist ropes 曳引绳
* hoistway 井道
hospital lift 医用电梯
hydraulic oil 液压油
** hydraulic elevator 液压电梯

☞ I

inner deck 内盖板
inspection 检查，检修
inspection button 检修按钮
** installation 安装
installation documents 安装文件
installation drawing 安装图纸
installation instructions 安装说明书
installation manager 安装经理
installation period 安装周期
installation personnel 安装人员
* installer 安装工
* instantaneous safety gear 瞬时式安全钳装置
ISO (international organization for standard) 国际标准化组织

☞ L

** landing 层站
** landing door 层门
* landing door jamp 层门门套
* landing indictor 层门位置显示装置
landing sills 层门地坎
* leveling 平层
leveling accuracy 平层准确度
** leveling device 平层装置
leveling error 平层误差
LS(limit switch) cam 限位开关碰铁
leveling inaccracy 平层误差
* leveling inductor plate 平层感应板
liftwell 电梯井道
** lift 电梯
* lift attendant 电梯司机
* limit switch 极限开关
load cell 称重装置
load measuring sensor 载重传感器
LS(limit switch) bracket 限位开关支架

☞ M

** machine room 机房
** machine-room-less elevator (MRL) 无机房电梯
machine room area 机房面积
machine room arrangement 机房布置
machine room depth 机房深度
machine room height 机房高度
* main drive chain guard 主驱动链保护装置
main floor 基站
main landing 基站
* maintenance man 维修人员
maintenance operation 维修作业
manufacture 制造，生产
manufacturer 生产者
mechanical safety shoe 机械式安全触板
metric 公制
motor 电动机，马达
motor base speed 电动机额定速度
manual opened door 手动门
motor capacity 电机功率
* moving walkway 自动人行道
multiple elevator hoistway 多电梯井道

☞ N

noise 噪声

☞ O

* observation lift 观光电梯
overload device (indicator)超载装置
* outer deck 外盖板
overspeed 超速
** overspeed governor 限速器

☞ P

** pallets 踏板
passenger car 载客轿厢
* passenger conveyor 自动人行道
** passenger lift 客梯
** pit 底坑
pit protection grid 底坑隔障
platform 轿厢底，轿底
phonic station reporter 语音报站
* protective earthing 保护接地
* permanent synchro motor 永磁电机
positive drive lift 强制驱动电梯
protective grating 防护栅栏
protective switch 保护开关
proximity protection device 近门保护装置
* progressive safety gear 渐进式安全钳
prototype 样机
pulse 脉冲

☞ R

** rated speed 额定速度
** rated load 额定载重量
rail 导轨
rail support 导轨支架
regular lay 交互捻 (绳)
relevel 再平层
re-leveling device 重平层装置
rope hitch 绳头组合
reliability 可靠性
residential lift 住宅电梯
* retrofit/ refurbish (电梯) 大修
* roller guide shoe 滚动导靴
* rope hitch plate 绳头板
* rope fastening 绳头组合
** rope gripper 夹绳器

☞ S

** safety gear 安全钳
* safety edge 安全触板
selector 选层器
* secondary sheave 复绕轮
shaft 井道
skirt guard safety device 裙板安全装置
* stairway 楼梯
standard 标准
** step 梯级
* step riser 梯级踢板
* step tread 梯级踏板
service lift 杂物电梯
service performance 维修作业
* sliding guide shoe 滑动导靴
* spring buffer 弹簧缓冲器
* step chain sprocket (wheel) 梯级链轮
step chain tension device 梯级链张紧装置
** skirt panel 裙板
storey 楼层
supporting beam 承重梁

☞ T

** talkback system 对讲系统

* tension pulley 张紧轮

toothed rack 齿条

** toe guard 护角板

traction 曳引，牵引

traction drive 曳引驱动

* traction lift 曳引电梯

** traction machine 曳引机

traction ratio 曳引比

* traction sheave 曳引轮

* travelling cable 随行电缆

** truss 桁架

two-speed sliding door 旁开门、双折门、双速门

two-to-four rope ratio 2:4 绕绳比

☞ U

U profile/ groove U形绳槽

U-bolt U形螺栓

up-down sloding door 垂直滑动门，上下滑动门

☞ V

* vibration absorber 减振器

vertically sliding door 垂直滑动门

vibration damper 减振器

☞ W

waiting queue 候梯乘客量

waiting signal 候梯信号

waiting time 候梯时间

wall of well 井道墙壁

* ward off rope device 挡绳装置

* weighing device 称重装置

* well 井道

well depth 井道深度

well enclosure 井道壁

worm 蜗轮

worm gear 蜗齿

worm gear reducer 蜗轮减速器

worm gearing 蜗轮传动装置

* worm reduction 蜗轮减速机

Appendix Ⅲ

Company Practical English
公司实用英语

缩写	全称	中文名称	类别
8D	eight disciplines	8原则/步骤	其他
AQL	acceptable quality level	允收品质水准	品保部
ASAP	as soon as possible	尽快	其他
ATA	actual time of arrival	实际到达时间	资材部
ATD	actual time of delivery	实际发货时间	资材部
AVL	approved vendor list	合格供应商名录	资材部
BOM	bill of material	物料清单	资材部
BTF	build to forecast	预测生产	资材部
BTO	build to order	接单生产	资材部
BTW	by the way	顺便说一句	其他
CAR	corrective action report	改善措施报告	报表
CC	carbon copy	抄送	邮件
CEO	chief executive officer	首席执行官	职位
CIP	continuous improvement plan	持续改善计划	报表
COD	cash on delivery	货到付款	资材部
CPK	process capability index	工序能力指数	报表
D/C	date code	日期代码	品保部
DMAIC	define/measure/analyze/improve/control	6sigma循环	品保部
DCC	document control center	文件管理中心	文控
DFMEA	design failure mode effect analysis	设计失效模式分析	其他
DIP	dual inline-pin package	双列直插式封装	制造部
DMM	digital multimeter	数字万用表	测试
DPPM	defect parts per million	百万中不良品数	工程部
ECN	engineering change notice	工程变更通知	其他
ECR	engineering change request	工程变更申请	报表
EE	electronic engineering	电子工程	工程部
ERP	enterprise resource planning	企业资源规划	其他
ESD	electrostatic discharge	静电放电	制造部
ETA	estimated time of arrival	预计到达时间	资材部
ETD	estimated time of delivery	预计发货时间	资材部
EVT	engineering verification testing	工程验证测试	测试
FA	failure analysis	不良分析	工程部
FAI	first article inspection report	首件检验报告	报表
FAQ	frequently asked questions	常见问题解答	其他
FAR	failure analysis report	不良分析报表	报表
FIFO	first in, first out	物料先进先出	资材部
FQC	Final quality control	终检品保	品保部
FYI	for your reference	供参考	其他
G R&R	gauge repeatability& reproducibility	重复和再现性	测试
HR	human resource dept.	人资部	部门
IE	industrial engineering	工业工程	工程部
IPQC	in-process quality control	制程品保	品保部
IQC	incoming(input) quality control	来料质量控制	品保部

缩写	全称	中文名称	类别
KPI	key performance index	关键绩效指标	工程部
L/T	lead time	生产周期	资材部
LCL	lower control limit	控制下限	报表
ME	mechanical engineering	机械工程	工程部
ME	manufacturing engineering	制造工程	工程部
MIS	management information system	资讯部	部门
MOQ	minimum order quantity	最小订购量	资材部
MRP	material requirement planning	物料需求计划	资材部
MSDS	material safety data sheet	物质安全数据表	品保部
NG	not good	不良	工程部
NPI	new product introduction	新产品导入	工程部
OA	office automation	办公自动化	其他
OQC	outgoing quality control	出货品保	品保部
P/N	part number	料号	品保部
PC	polycarbonate	聚碳酸酯	塑料
PCB	printed circuit board	印刷电路板	制造部
PDCA	plan-do-check-action	戴明循环	品保部
PE	product engineering	产品工程	工程部
PM	project management	专案管理	品保部
PMP	process management plan	制程管理计划	制造部
PnP	plug and play	即插即用	其他
PO	purchase order	订单	资材部
QA	quality assurance dept.	品保部	部门
QC	quality control dept.	质检部	部门
QE	quality engineering	品质工程	工程部
Qty	quantity	数量	制造部
R/C	root cause	真正原因	工程部
RD	research and development dept.	研发部	部门
rev.	revision	版本	文控
RFQ	request for quotation	询价	资材部
RH	relative humidity	相对湿度	资材部
ROHS	restriction of hazardous substance	有害物质限制	品保部
S/N	serial number	序列号	其他
SA	sales department	业务部	部门
SIP	standard inspection procedure	检验标准/检规	品保部
SMT	surface mount technology	表面贴装技术	制造部
SOP	standard operation procedure	操作标准	品保部
SPC	statistical process control	统计制程管制	报表
SQM	supplier quality management	供应商品质管理	品保部
TE	test engineering	测试工程	工程部
TQM	total quality management	全面品质管理	品保部
TT	telegraphic transfer	电汇	资材部
UPS	uninterrupted power supply	不间断电源	其他

缩写	全称	中文名称	类别
USL	upper specification limit	规格上限	报表
VMI	vendor management inventory	供应商管理库存	资材部
VP	vice president	副总裁	职位
W/H	warehouse	仓库	资材部
WI	working instruction	工作指导书	制造部
	approval sheet	承认书	品保部
	ISO9000	质量管理体系标准	品保部
	ISO14000	环境管理体系标准	品保部
	bar code	条形码	制造部
	gross weight	毛重	制造部
	net volume / content	净含量	制造部
	tape reel	载带	制造部
	tray	托盘	制造部
	carton	硬纸箱	制造部
	packing specification	包装规范	制造部
	antistatic coat	静电衣	制造部
	magnifier	放大镜	制造部
	microscope	显微镜	制造部
	reflow oven / furnace	回焊炉	制造部
	wave soldering machine	波峰焊设备	制造部
	cavity No.	穴号	制造部
	mold No.	模号	制造部
	trouble shooting	检修	制造部
	work order	工单	制造部
	NG rate	不良率	制造部
	day shift	白班	制造部
	semi-finished product	半成品	制造部
	trial run	试产	制造部
	logistics management	物流管理	资材部
	purchasing Division	采购科	资材部
	quotation	报价	资材部
	lot. No.	批号	资材部
	hand carry	直接随身携带	资材部
	express	快递	资材部
	vendor hub	供应商仓库	资材部
	invoice No.	发票号码	资材部
	packing list	装箱单	资材部
	forklift	叉车	资材部
	projector	投影仪	办公
	printer	打印机	办公
	scanner	扫描仪	办公
	meeting room	会议室	办公
	stapler	订书机	办公

缩写	全称	中文名称	类别
	file	文件夹	办公
	correction tape	修正带	办公
	official stamp	公章	办公
	check list	点检表/核对表	报表
	maximum value	最大值	报表
	average value	平均值	报表
	threshold / critical value	临界值	报表
	histogram	柱状图	报表
	pareto chart / diagram	柏拉图	报表
	conference call	电话会议	会议
	minutes	会议记录	会议
	preside (over)	主持会议	会议
	meeting room	会议室	会议
	general manager	总经理	职位
	manager	经理	职位
	engineer	工程师	职位
	technician	技术员	职位
	operator	操作员	职位
	adjustor	调试人员	职位
	inspector	检验员	职位
	assistant	助理	职位
	clerk	文员	职位
	top management	高层管理者	职位
	factory manager	厂长	职位
	workshop directory	车间主任	职位
	colleague	同事	职位
	supervisor	主管	职位
	address	地址	邮件
	zip code / postcode	邮编	邮件
	mobile phone / cell phone	移动电话	邮件
	manufacturing dept.	制造部	部门
	material dept.	资材部	部门
	engineering dept.	工程部	部门
	equipment dept.	设备部	部门
	marketing dept.	市场部	部门
	customer service dept.	客服部	部门
	planning dept.	企划部	部门
	general affairs dept.	总务部	部门
	rework	重工	工程部
	sorting	挑选	工程部
	hold	暂放	工程部
	burn-in	烧机测试	测试

参考文献

References

[1] 黄星，夏玉波. 机电一体化专业英语. 北京：人民邮电出版社，2010.

[2] 肖伟平，谢英星. 模具专业英语. 第2版. 大连：大连理工大学出版社，2009.

[3] 江华圣. 电工电子专业英语. 第2版. 北京：人民邮电出版社，2010.

[4] 金利. 职场英语. 求职你准备好了吗. 西安：世界图书出版西安公司，2010.

[5] 李庆芬. 机电工程专业英语. 哈尔滨：哈尔滨工程大学出版社，2007.

[6] 赵运才，何法江. 机电工程专业英语. 北京：北京大学出版社，2006.

[7] 贺德明，肖伟平. 电梯结构与原理. 广州：中山大学出版社，2009.

[8] 李折昆. 机电工程英语. 武汉：华中科技大学出版社，2008.

[9] 王会，吕江毅. 汽车专业英语. 北京：化学工业出版社，2010.

[10] 刘建雄，王家惠，廖丕博. 模具设计与制造专业英语. 北京：北京大学出版社，2006.

[11] 石定乐，孙嫘. 汽车制造英语. 武汉：武汉大学出版社，2010.

[12] 刘小芹，刘骋. 电子与通信技术专业英语. 北京：人民邮电出版社，2008.

[13] 廖伟. 电子行业实用英语. 第2版. 北京：世界图书出版公司，2006.

[14] Installation Manual：Basic Field Practices for Installation of Elevators and Escalators.Published by Elevator World，Inc.

[15] George Strakosch.The Vertical Transportation Hand book,4th ed. Copyright John Wiley &Sons.Inc.,2010.

[16] A17.1 Safety Code for Elevators and Escalators, copyright American Society of Mechanical Engineers.

[17] Elevator World March 1999, November 1999 and April 2001.